The Periodic Table at a Glance

The Periodic Table at a Glance

Dr M. A. Beckett

Senior Lecturer in Inorganic Chemistry,
University of Wales, Bangor

Dr A. W. G. Platt

Senior Lecturer in Forensic Science,
Staffordshire University

Blackwell
Publishing

© 2006 M.A. Beckett and A.W.G. Platt

Blackwell Publishing Ltd,
Editorial offices:
Blackwell Publishing Ltd, 9600 Garsington Road, Oxford OX4 2DQ, UK
    Tel:+44 (0)1865 776868
Blackwell Publishing Inc., 350 Main Street, Malden, MA 02148-5020, USA
    Tel:+1 781 388 8250
Blackwell Publishing Asia Pty Ltd, 550 Swanston Street, Carlton, Victoria 3053, Australia
    Tel:+61 (0)3 8359 1011

First published 2006 by Blackwell Publishing Ltd

ISBN-13: 978-14051-3299-2
ISBN-10: 1-4051-3299-X

Library of Congress Cataloging-in-Publication Data
Beckett, M. A. (Mike A.)
    The periodic table at a glance / M.A. Beckett, A.W.G. Platt.– 1st ed.
        p. cm.
    Includes index.
    ISBN-13: 978-1-4051-3299-2 (pbk. : alk. paper)
    ISBN-10: 1-4051-3299-X (pbk. : alk. paper)   1. Periodic law–Tables–Juvenile
literature.   I. Platt, A.W.G. (Andy W. G.)   II. Title.

    QD467.B38   2006
    546′.8–dc22

                                                    2006001692

A catalogue record for this title is available from the British Library

Set in 9.5/11.5 pt Times
by TechBooks, New Delhi, India
Printed and bound in Great Britain
by TJ International, Padstow, Cornwall

The publisher's policy is to use permanent paper from mills that operate a sustainable forestry policy, and which has been manufactured from pulp processed using acid-free and elementary chlorine-free practices. Furthermore, the publisher ensures that the text paper and cover board used have met acceptable environmental accreditation standards.

For further information on Blackwell Publishing, visit our website:
www.blackwellpublishing.com

Contents

Contents

Introduction

This book has been written primarily as a revision guide to assist undergraduate students in their study of degree level introductory inorganic chemistry (1st and 2nd year). We hope that it will not only be of value to students reading chemistry as their main degree subject, but that it will prove useful to students reading other science degrees or those in which chemistry forms a significant part. The layout of the book is based on the arrangement of the Periodic Table, and information is presented in 'bite-size' double page spreads covering, in the authors' experiences, all the essential topics. The chemical reactivity of the elements and their compounds is described in relation to the position of the element within the Periodic Table, with all the necessary linking underlying concepts and principles appropriately summarised. Important industrial processes and related chemistry are also presented. A selection of more detailed texts is given in the 'Suggested Further Reading' section at the end of this book.

Acknowledgements

We are grateful to all of those who have helped with the preparation of the book, especially Catherine Bland for drawing some of the diagrams. To all of these we record our sincere thanks.

We would like to dedicate this book to Lindy Beckett and Dr Janet Wedgwood for their continuous support and encouragement.

Atoms, the Periodic Table and Periodic Properties

1.1. Atomic Structure (I)

Introduction

The 'classical' description of an atom is one in which negatively charged electrons orbit around a positively charged nucleus comprising protons, which bear the positive charge, and neutrons. The number of protons in the nucleus is referred to as the **atomic number**, **Z**, of the element. The number of protons (and hence the number of electrons in the neutral atom) defines the chemical identity of the atom. All elements with a single proton in the nucleus are hydrogen regardless of how many neutrons are present. Similarly with two protons in the nucleus we have helium and with 92 protons uranium, and so on. The number of electrons around a nucleus depends on the chemical environment of the atom and can change; the number of protons in the nucleus cannot be changed by chemical processes. Chemical processes are intimately linked with the movement of electrons between atoms. Thus an appreciation of the properties of electrons within atoms and molecules is fundamental to an understanding of chemistry.

Electronic properties and wavefunctions

Modern understanding of the electronic structure of atoms is derived from a number of fundamental findings on the properties of matter at an atomic level. The behaviour at this scale is very different from that which we are used to in everyday life, and as the laws of classical mechanics do not apply, such behaviour is by no means intuitive.

It was found that electrons within an atom can only have certain well-defined energies. This **quantisation of energy** means that electrons can only absorb specific amounts of energy and in doing so undergo transitions from one allowed energy level to another (see Figure 1.1a). Energy, in the form of radiation, can only be absorbed or emitted in discrete quanta, and the relationship between this energy ΔE (units J), and the frequency of radiation v (units Hz or s^{-1}) is:

$$\Delta E = h v$$

where h is Planck's constant 6.626×10^{-34} Js and ΔE is the difference between two allowed values of the energy of the electron.

Matter at the atomic level has properties normally associated with electromagnetic radiation, i.e. electrons behave as waves as well as particles. This is known as the **wave–particle duality**. The relationship between the classical property, momentum, p and the wave characteristic, wavelength, λ (units m) is:

$$p = mv = h/\lambda$$

where m and v are the mass (kg) and velocity (ms^{-1}) of the particle. In contrast to classical mechanical systems it is not possible to simultaneously determine the energy and position of a particle. This finding, the **Heisenberg Uncertainty Principle**, is summarised by the equation:

$$\Delta p \Delta x \geq h/2\pi$$

where p is the momentum of the particle and x its position. The term Δ means the uncertainty in a given quantity, thus Δp is the uncertainty in the momentum (and hence energy). Thus instead of the system of electrons orbiting a nucleus having precisely defined energy, we have to deal with **probable** values of energy and position of the electrons.

2

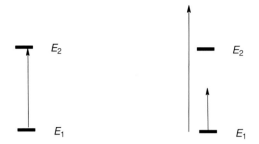

If the energy difference between the two states is $\Delta E = E_2 - E_1$ then only when the energy supplied exactly matches this (left hand side diagram) will absorption occur. In the situations illustrated by the right hand diagram absorption will not occur.

Figure 1.1a   Allowed transitions between electronic energy levels.

Wave mechanics is the mathematical treatment of the properties of the electron and leads to many important results which chemists now routinely use to explain structures and reactivity, without having to encounter the mathematical details of the solutions of the equations. An equation to describe the motion of the electron in the electric field of a nucleus was devised by Schrödinger (Figure 1.1b). This equation has an infinite number of possible solutions; these solutions being values of total energy, E and **wavefunctions**, ψ. The wavefunctions themselves are simply mathematical functions that describe the volume of space occupied by the electron, i.e. the **atomic orbital**, its shape and symmetry properties. Figure 1.1c shows the plot of the wavefunction of the 3s orbital $\psi_{3s} = k(ar - r^2)e^{-br}$, where r is the distance of the electron from the nucleus and k, a and b are collections of fundamental constants. The **electron density** distribution is given by ψ^2 and gives the **probability** of finding the electron at any given distance from the nucleus.

$$\frac{-h^2\nabla^2\psi}{8\pi^2 m} - \frac{ze^2\psi}{4\pi\varepsilon_0 r} = E\psi$$

or kinetic energy + potential energy = total energy

ψ is the wavefunction of the electron, e its charge, m its mass and r its distance from the nucleus; z is the charge on the nucleus, ε_0 is the permittivity of a vacuum (a measure of its electrical insulation properties) and E is the energy of the electron.

Figure 1.1b   The Schrödinger equation.

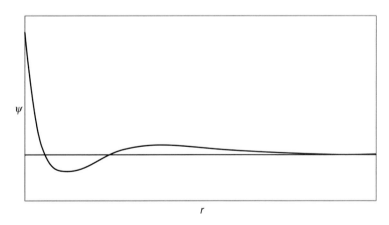

Figure 1.1c   The radial wavefunction for the 3s orbital.

1.2. Atomic Structure (II)

Quantum numbers

The solutions of the Schrödinger equation led to the properties of the electrons being expressed as a series of **quantum numbers**. These quantum numbers govern the energy, symmetry and number of any given type of atomic orbital. Each electron is uniquely defined by four quantum numbers, n, l, m_l and m_s. These are hierarchical with the value of one determining which values the others can adopt, as outlined below.

The **principal quantum number**, n, governs the energy of the electron and can take any integer value from 1 upwards. For the first period $n = 1$, for the second $n = 2$, but after this, the apparent simplicity is lost as discussed later in this section. Next in the hierarchy is the **orbital quantum number**, l. This can take integer values from zero up to $n - 1$. Thus for $n = 1$ the only possible value for l is $l = 0$. For $n = 2$, l can have values of 0 or 1, and so on. The numerical value of l defines the type of orbital and to a lesser extent than n, its energy. There are only four chemically significant values of l and hence four types of atomic orbital. These are s-orbitals ($l = 0$), p-orbitals ($l = 1$), d-orbitals ($l = 2$) and f-orbitals ($l = 3$). The shapes of the s-, p- and d-orbitals are shown in Figure 1.2a.

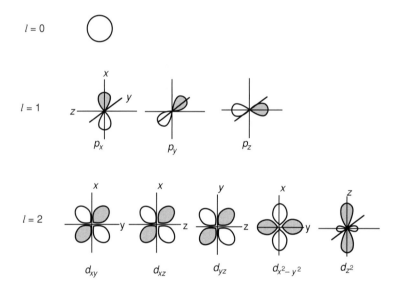

Figure 1.2a The shapes and numbers of the $1s$-, $2p$- and $3d$-orbitals.

The **magnetic quantum number**, m_l, has allowed integer values from $-l$ to $+l$, governs the behaviour of electrons in a magnetic field and also gives us the number of any one type of orbital (a single s-orbital, $3 \times p$, $5 \times d$ and $7 \times f$-orbitals). Thus for $l = 0$ (an s-orbital) m_l can only have one value, $m_l = 0$, and hence there is only one s-orbital for each principal quantum number. When $l = 1$ (a p-orbital) m_l can have values of -1, 0 and $+1$. As there are three values it means that there are three p-orbitals for each principal quantum number. The same arguments explain why there are five d-orbitals and seven f-orbitals for each principal quantum number for which d- and f-orbitals are allowed.

For each combination of n, l and m_l there are two allowed values of the **spin quantum number**, m_s. This describes the 'spin' properties of the electron and can take values of $+^1/_2$ or $-^1/_2$. Thus, each atomic orbital defined by its n, l and m_l values can hold two electrons: one with $m_s = +^1/_2$ and one with $m_s = -^1/_2$.

Using these quantum numbers it is possible to explain the structure of the **Periodic Table**. Figure 1.2b illustrates how for $n = 1$ and 2 we expect to find two and eight unique sets of quantum numbers, and hence two and eight elements in rows 1 and 2 respectively. This matches exactly with the observed structure. For the third row, where $n = 3$, we would expect to find 18 elements, but this is not the case. The reason for this is that whilst the principal quantum number has the major influence on the energy of the orbital, l also has a significant effect. Figure 1.2c shows a qualitative energy level diagram for the H atom and non-H atoms. For hydrogen all orbitals with the same principal quantum number have the same energy. In non-H atoms the order of energies is $ns < np < nd < nf$, with the lowest energy orbitals filled first, the '**aufbau**' (**or build up) principle**. The ordering of the nd versus $(n + 1)s$ depends on Z. After the $3p$ level we would have expected the $3d$ to be filled next, but it turns out that for $Z = 19$ (K) and 20 (Ca) the $4s$ has a lower energy than the $3d$ and it is occupied in preference. Thus the third period has only eight elements. After $Z = 21$ (Sc) the $3d$-orbitals have similar energy to the $4s$ and both $4s$ and $3d$

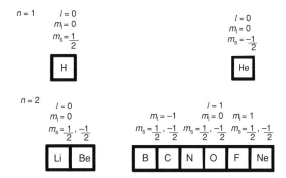

Figure 1.2b How quantum numbers explain the arrangement of the first two rows of the periodic table. The assignment of values of m_l and m_s to particular elements is arbitrary.

are occupied in the ground state of the d-block metals. By $Z = 30$ (Zn) the $3d$ is much lower than the $4s$ and is essentially a **core orbital** for this and subsequent elements. Hence for the elements Ga to Kr the valence orbitals are the $4s$ and $4p$. For the same reasons as outlined above the $4f$ is not of sufficiently low energy to be occupied until period 6.

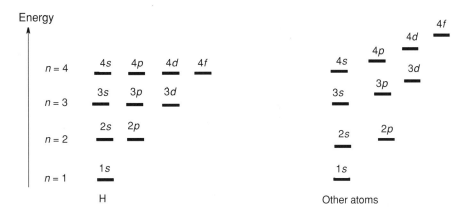

Figure 1.2c Orbital energies of H and non-H atoms.

Core and valence orbitals

Orbital energies are also affected by the charge on the nucleus; the higher the nuclear charge the greater the attraction between the electron and the nucleus and hence the lower its energy. Thus on increasing atomic number the orbital energies decrease. This means that not all the electrons around an atom are available for chemical processes. Those at very low energy are too stable and are termed **core electrons**. The occupied orbitals of highest energy in an atom, the so-called **valence orbitals**, contain electrons that are the least strongly held. These can be transferred to other atoms and it is only these that are involved in the formation of chemical bonds. The size of atomic orbitals increases with the principal quantum number for a given element.

1.3. The Periodic Table

Overview

The Periodic Table is a cornerstone of inorganic chemistry, and as such is the secure foundation upon which this book is built. In Section 1.2 the electronic configurations of the elements were obtained by applying the aufbau principle, Figure 1.3a. The Periodic Table is a pictorial representation of the outcome of this procedure, and as such there are many ways in which it can be drawn. A frequently used version is shown in the inside cover, and this is the one adopted for this book. The data presented in this **Periodic Table** includes the atomic number of the element, its name, its chemical symbol, and its relative atomic mass. In Figure 1.3b the symbols and the names of the elements are given in alphabetical order.

$n + l$				
1	$1s$			
2	$2s$			
3	$2p$	$3s$		
4	$3p$	$4s$		
5	$3d$	$4p$	$5s$	
6	$4d$	$5p$	$6s$	
7	$4f$	$5d$	$6p$	$7s$
8	$5f$	$6d$		

Figure 1.3a The sequence of filling of atomic orbitals (the aufbau principle). The arrangement is determined by the value of $n + l$ of any particular orbital, with orbitals with lowest $n + l$ values filling first. When there are several orbitals with equivalent $n + l$ values then the orbital with lowest n value fills first.

Actinium Ac 89	Aluminium Al 13	Americium Am 95	Antimony Sb 51
Argon Ar 18	Arsenic As 33	Astatine At 85	Barium Ba 56
Berkelium Bk 97	Beryllium Be 4	Bismuth Bi 83	Bohrium Bh 107
Boron B 5	Bromine Br 35	Cadmium Cd 48	Caesium Cs 55
Calcium Ca 20	Californium Cf 98	Carbon C 6	Cerium Ce 58
Chlorine Cl 17	Chromium Cr 24	Cobalt Co 27	Copper Cu 29
Curium Cm 96	Dubnium Db 105	Dysprosium Dy 66	Einsteinium Es 99
Erbium Er 68	Europium Eu 63	Fermium Fm 100	Fluorine F 9
Francium Fr 87	Gadolinium Gd 64	Gallium Ga 31	Germanium Ge 32
Gold Au 79	Hafnium Hf 72	Hassium Hs 108	Helium He 2
Holmium Ho 67	Hydrogen H 1	Indium In 49	Iodine I 53
Iridium Ir 77	Iron Fe 26	Krypton Kr 36	Lanthanum La 57
Lawrencium Lr 103	Lead Pb 82	Lithium Li 3	Lutetium Lu 71
Magnesium Mg 12	Manganese Mn 25	Meitnerium Mt 109	Mendelevium Md 101
Mercury Hg 80	Molybdenum Mo 42	Neodymium Nd 60	Neon Ne 10
Neptunium Np 93	Nickel Ni 28	Niobium Nb 41	Nitrogen N 7
Nobelium No 102	Osmium Os 76	Oxygen O 8	Palladium Pd 46
Phosphorus P 15	Platinum Pt 78	Plutonium Pu 94	Polonium Po 84
Potassium K 19	Praseodymium Pr 59	Proctactinium Pr 91	Promethium Pm 61
Radium Ra 88	Radon Rn 86	Rhenium Re 75	Rhodium Rh 45
Rubidium Rb 37	Ruthenium Ru 44	Rutherfordium Rf 104	Samarium Sm 62
Scandium Sc 21	Seaborgium Sg 106	Selenium Se 34	Silicon Si 14
Silver Ag 47	Sodium Na 11	Strontium Sr 38	Sulfur S 16
Tantalum Ta 73	Technetium Tc 43	Tellurium Te 52	Terbium Tb 65
Thallium Tl 81	Thorium Th 90	Thulium Tm 69	Tin Sn 50
Titanium Ti 22	Tungsten W 74	Uranium U 92	Vanadium V 23
Xenon Xe 54	Ytterbium Yb 70	Yttrium Y 39	Zinc Zn 30
Zirconium Zr 40			

Figure 1.3b Alphabetical listing of the elements with symbol and atomic number (Z).

Atomic masses and atomic numbers

Classical atomic theory defines an element in terms of its **atomic number** (Z) but the nucleus may also contain uncharged neutrons. A neutron has a similar mass to that of the proton ($\sim 10^{-27}$ kg), and the mass of an electron ($\sim 10^{-30}$ kg) is negligible in comparison. Since the mass of an atom is extremely small chemists prefer to use the **atomic mass unit** (amu) where 1 amu is defined as the mass of one proton. An element's relative atomic mass may vary depending upon how many neutrons (N) there are in its nucleus. In practice, only certain combinations are stable (naturally abundant) and the number of neutrons in the nucleus is generally equal to, or slightly greater than, the number of protons in the nucleus. The sum of the number of protons and the number of neutrons ($Z + N$) is called the **mass number** (A). The term *nuclide* refers to a specific nucleus with a defined number of protons and neutrons, e.g. $^{11}_{5}$B (or more simply, ^{11}B, since all nuclei with five protons in them are by definition B nuclei). **Isotopes** are defined as nuclides with the same atomic number (same element) but different neutron numbers.

> Mass number = number of protons + number of neutrons, or $A = Z + N$
> Nuclides are represented as $^{A}_{Z}$E or ^{A}E, where E = the symbol for the element

Average **relative atomic masses** are used for the elements as this represents a weighted average of all the naturally occurring isotopes of the particular element. Thus for B we see that the relative atomic mass is 10.81. This originates from the fact that $\sim 80\%$ of naturally abundant boron nuclei are $^{11}_{5}$B and $\sim 20\%$ are $^{10}_{5}$B.

The Periodic Table

The elements in the Periodic Table shown on the inside cover are arranged in increasing atomic number. The principal aim of the Periodic Table is to display the elements in such a way that gives an informed insight into their electronic structures. This is important since the chemistry of an element is dependent upon how its outermost (valence) electrons interact with valence electrons of other elements.

> The elements are arranged in rows or *periods* (numbered 1–7), *blocks* (described as the s-, p-, d-, f-block), and columns or *groups* (numbered 1–18).

The number of elements within a **period** of the Periodic Table varies from 2 to 32: the first period contains just 2 elements (H and He); the second period 8 elements (Li to Ne); the third period 8 elements (Na to Ar); the fourth period 18 elements (K to Kr); the fifth period 18 elements (Rb to Xe); the sixth period 32 elements (Cs to Rn, and includes La to Yb); the seventh period the remaining 23 known elements (Fr to the heaviest known element, element 109, and includes Ac to No). The **period numbers** are derived from the **principal quantum number** (n) of the single valence electron of the alkali metal contained within that period. It is effectively the first electron within that particular electron shell. Upon moving across the Periodic Table the valence shell starts to fill up.

The various **blocks** are labelled the s-, p-, d- or f-block and they take their name from the **orbital quantum number** ($l = 0, 1, 2, 3$, respectively) of valence orbitals that are being filled according to the aufbau principle. The number of elements permitted within each row of each block is determined by the **magnetic quantum number** (m_l) which in turn is related to l and m_l (i.e. 2 for s, 6 for p, 10 for d, and 14 for f). By inspection of the Periodic Table it should be possible to obtain the ground state electronic configuration of any element. The **groups** of the Periodic Table are numbered 1 to 18, and are used to differentiate between the various columns within the blocks.

Historical perspective

A Periodic Table of the elements, in a pattern not too different from that used today, was first described as early as 1869 by the Russian scientist Mendeleyev. He based his table on relative atomic masses and proposed in his so-called periodic law (the properties of the elements are a periodic function of their atomic weights) that elements with similar chemical and physical properties (e.g. F, Cl, Br, I) should be grouped together as a related family. Some elements had yet to be discovered (e.g. Sc, Ga, Ge). Mendeleyev described their expected properties, left gaps in his Periodic Table and awaited their future discovery. These elements were soon isolated but an explanation for the observed pattern of the Periodic Table had to await, initially, Bohr's model of the atom (1913), and, finally, the quantum numbers and atomic orbitals derived from Schrödinger's wave equations (1926).

1.4. Periodic Properties

Atomic and ionic radii

The dimensions of species such as atoms and ions have an important bearing on their chemical behaviour.

Atomic radius is defined as either half the internuclear distance in metals (**metallic radius**) or half the internuclear distance between singly bonded atoms of the same element in a molecule (**covalent radius**). In practice, the values quoted for covalent radii are averages from a number of determinations over a range of molecules. There is a decrease in atomic radius in moving across any period of the Periodic Table. This is due to the increase in nuclear charge (from 3 to 9 going from Li to F). The additional electrons (in the same shell) do not shield each other effectively from this increased charge. Thus they experience an increased **effective nuclear charge** and are therefore held closer to the nucleus. The decreases from Group 1 to Group 18 are 145 pm to 50 pm (Li–F), 180 to 100 pm (Na–Cl), 220 to 115 pm (K–Br), and 235 to 140 pm (Rb–I). On descending any group there is an increase in radius as expected from the extra shell of electrons. The usefulness of atomic/covalent radii is that they give good estimates for the internuclear distances expected in chemical compounds. Thus, since the covalent radii of carbon and nitrogen are 77 and 75 pm respectively, then a $C-N$ single bond would be expected to have a bond length of 152 pm. A bond distance significantly smaller than this would imply the existence of an additional bonding interaction, for example, a double or triple bond.

Ionic radii are derived from the separation of ions in crystal lattices (Section 2.4). Experimentally this is determined by X-ray diffraction. This technique locates regions of high electron density and thus gives internuclear separations (as the electron density is highest in the region of the nucleus). Cations, which have fewer electrons, are **always smaller** than the parent atom. The ionic radius quoted for an element depends on its co-ordination environment. For instance Li^+ has ionic radii of 73 pm (4 co-ordinate), 90 pm (6 co-ordinate) and 106 pm (8 co-ordinate). The reason for this measured increase with co-ordination number is that with more groups around the Li^+ there are increased repulsions and thus anions move further away to reduce this, leading to an apparent increase in the size of the cation. Cation sizes decrease with increased charge. Thus in the series of isoelectronic ions Na^+, Mg^{2+}, Al^{3+}, the six co-ordinate ionic radii are 116, 86 and 68 pm respectively. Similarly, anions are **always larger** than their parent atoms due to the increased electron–electron repulsion on adding extra electrons; for example S^{2-} (184 pm) is larger than Cl^- (167 pm). The **van der Waals radius** is the distance of closest approach between two atoms before there is any bonding interaction. Thus if two atoms are found to be closer together than the sum of their van der Waals radii, then there is some bonding interaction between them.

Ionisation energies, electron affinities and electronegativities

Ionisation energies (see Figure 1.4a) are important as they give a measure of the ease with which electrons can be **removed** from an atom or ion, and thus their propensity to be involved in chemical bonding. They are not a complete description, of course, but give a useful guide. The general trends across and down a period are as might be expected from atomic size, and some of these can be seen from the data in Figure 1.4b. As the atom becomes smaller it implies that the outer electrons are more firmly bound to the atom and hence more difficult to remove (ionise). Thus moving across a row there is a general increase in ionisation energy. There are, however, important discontinuities in this trend where particularly **stable electronic configurations** are involved. For example the stable $2s^2$ and $2p^3$ configurations of beryllium and nitrogen respectively mean that their first ionisation energies are higher than those of boron and oxygen. On descending a group the general trend is the

$$E_{(g)} \rightarrow E^+_{(g)} + e^-_{(g)} \qquad \text{First ionisation energy}$$

$$E^+_{(g)} \rightarrow E^{2+}_{(g)} + e^-_{(g)} \qquad \text{Second ionisation energy}$$

$$E_{(g)} + e^-_{(g)} \rightarrow E^-_{(g)} \qquad \text{First electron affinity}$$

$$E^-_{(g)} + e^-_{(g)} \rightarrow E^{2-}_{(g)} \qquad \text{Second electron affinity}$$

Figure 1.4a Equations defining ionisation energies and electron affinities.

Li	Be	B	C	N	O	F	Ne
520	900	801	1086	1402	1314	1681	2081
(−59.8)	(−36.9)	(−17.6)	(−153.3)	(−7.0)	(−141.3)	(−328.0)	(28.9)
Na	Mg	Al	Si	P	S	Cl	Ar
496	738	578	787	1012	1000	1251	1521
(−52.5)	(185.9)	(−19.6)	(−131.0)	(−70.2)	(−196.8)	(−349.2)	(35.7)
K	Ca	Ga	Ge	As	Se	Br	Kr
419	589	579	762	947	941	1140	1351
(−48.4)	(186.0)	(−28.9)	(−119)	(−78)	(−195.0)	(−324.6)	(40.5)
Rb	Sr	In	Sn	S	Te	I	Xe
403	550	558	708	831	869	1008	1170
(−46.9)	(145)	(−28.9)	(−107.3)	(−103.2)	(−190.2)	(−295.2)	(43.5)
Cs	Ba	Tl	Pb	Bi	Po	At	Rn
376	503	589	716	703	812	930	1037
(−45.5)	(46.0)	(−19.2)	(−35.1)	(−91.2)	(−183.3)	(−270.1)	

Figure 1.4b First ionisation energies of the s- and p-block elements. Note that the values are always endothermic as energy is required to remove electrons from a neutral atom, and that the general trend of a decrease in ionisation energy on descending a group is not always followed for the p-block. The first electron affinities are given in parentheses and are generally mildly exothermic. All values are in $kJmol^{-1}$.

expected decrease in first ionisation energy with increasing atomic size (but see important differences in the higher ionisations for p-block elements in Section 4.1).

Electron affinity is a measure of the ease with which electrons can be **added** to atoms or ions. It is of major importance only to elements at the right hand side of the Periodic Table as on gaining one or two electrons, these atoms can achieve stable electronic configurations. The first electron affinities are often exothermic, whilst the addition of further electrons is **always** endothermic due to increased electron–electron repulsion. For instance, $O + e^- \rightarrow O^-$ $\Delta H = -141 \, kJmol^{-1}$, but for $O^- + e^- \rightarrow O^{2-}$ $\Delta H = +844 \, kJmol^{-1}$, despite achieving an octet configuration. Some electron affinities are given in Figure 1.4b – note that values can be either exothermic or endothermic in contrast to ionisation energies which are always endothermic.

Electronegativity is a quantity related to the ability of an atom within a chemical compound to attract electrons to itself. There are several scales which give numerical values calculated by different methods, but the general trends are the same for each. Some data for the Allred–Rochow scale are given in Figure 1.4c. Values of high electronegativity (~4 for F being the highest) indicate that, in a bond, F attracts electron density towards itself and will have a residual partial negative charge. The highest values for electronegativity are found for the elements at the extreme upper right of the Periodic Table, whilst the lowest (~ 1 for Cs) are for the metals at the lower left. Thus the C−F bond where two electrons are shared to form the covalent bond would be **polar**, with residual positive and negative charges residing on the carbon and fluorine atoms respectively. This is represented as $^{\delta+}C-F^{\delta-}$. In general, for any bond between dissimilar elements the resulting bond will be polar. At the extremes, bonding between elements of vastly different electronegativities leads to ionic bonds.

Li	Be	B	C	N	O	F
0.97	1.47	2.01	2.50	3.07	3.50	4.10
Cs	Ba	Tl	Pb	Bi	Po	At
0.86	0.97	1.44	1.55	1.67	1.76	1.90

Figure 1.4c Allred–Rochow electronegativities for selected s- and p-block elements.

1.5. Redox Processes (I)

Standard reduction potentials

Almost all elements can exist in more than one oxidation state. A useful measure of the ease of conversion of an element or compound **in solution** from one oxidation state to another is the **standard reduction potential**, E^0. Tabulated values generally refer to aqueous solution, and data for other solvents is more limited. The reduction potential is the difference in electric potential between an oxidised and reduced form of an element and is represented by a **half equation**, for example:

$$Fe^{3+}(aq) + e^- \rightleftharpoons Fe^{2+}(aq) \quad E^0 = 0.77\,V$$

Note that this is written as a reduction (of iron(III) to iron(II)) which is the convention adopted. To represent an oxidation of iron(II) to iron(III) we simply reverse the equation and change the sign of the potential:

$$Fe^{2+}(aq) \rightleftharpoons Fe^{3+}(aq) + e^- \quad E^0 = -0.77\,V$$

The sign and magnitude of the potential indicate the inherent ease of the process. As the Fe^{3+}/Fe^{2+} potential is positive and reasonably high we can say that the process is an inherently favourable one, i.e. Fe^{3+} is a moderate oxidising agent. For reduction potentials the range of values is rather compressed, with values between $\pm 3V$ representing the extremes as shown in Figure 1.5a. At the top of Figure 1.5a we have strong oxidising agents such as F_2 on the left hand side of the half equation and weak reducing agents such as F^- on the right hand side, i.e. the 2.87 V tells us that the reduction of F_2 to F^- is favourable, whilst the reverse reaction, the oxidation of F^- to F_2, is very difficult to bring about. Similarly we can see that the Na^+ ion is a poor oxidant whilst Na metal is a very strong reducing agent. To represent chemical reactions we must combine two half equations in the appropriate manner, i.e. make one half represent a reduction and the other half represent an oxidation. An example is shown in Figure 1.5b. The sum of the potentials for the two halves of the reaction is called the **cell potential** and gives an indication of the spontaneity of the reaction, a positive value meaning that the reaction should proceed. The underlying thermodynamics is expressed in the equation:

$$\Delta G^0 = -nFE^0_{cell}$$

where n is the number of electrons transferred and F is the Faraday constant.

Thus a positive value of E^0_{cell} will give a decrease in free energy, $\Delta G^0 < 0$, and hence the process will be spontaneous.

F_2	$+ 2e^-$		$\rightleftharpoons$	$2F^-$		$E^0 = 2.87\,V$
MnO_4^-	$+ 8H^+$	$+ 5e^-$	$\rightleftharpoons$	Mn^{2+}	$+ 4H_2O$	$E^0 = 1.52\,V$
O_2	$+ 4H^+$	$+ 4e^-$	$\rightleftharpoons$	$2H_2O$		$E^0 = 1.23\,V$
$Cr_2O_7^{2-}$	$+ 14H^+$	$+ 6e^-$	$\rightleftharpoons$	$2Cr^{3+}$	$+ 7H_2O$	$E^0 = 1.33\,V$
Cu^{2+}	$+ 2e^-$		$\rightleftharpoons$	Cu		$E^0 = 0.34\,V$
$2H^+$	$+ 2e^-$		$\rightleftharpoons$	H_2		$E^0 = 0.00\,V$
CO_2	$+ 2H^+$	$+ 2e^-$	$\rightleftharpoons$	HCO_2H		$E^0 = -0.20\,V$
H_3PO_3	$+ 2H^+$	$+ 2e^-$	$\rightleftharpoons$	H_3PO_2	$+ H_2O$	$E^0 = -0.56\,V$
Zn^{2+}	$+ 2e^-$		$\rightleftharpoons$	Zn		$E^0 = -0.76\,V$
Al^{3+}	$+ 3e^-$		$\rightleftharpoons$	Al		$E^0 = -1.68\,V$
Na^+	$+ e^-$		$\rightleftharpoons$	Na		$E^0 = -2.71\,V$
$3N_2$	$+ 2H^+$	$+ 2e^-$	$\rightleftharpoons$	$2HN_3$		$E^0 = -3.10\,V$

Figure 1.5a Some selected reduction potentials. Note that the oxidant is always on the left hand side of the equation and the reducing agent on the right hand side.

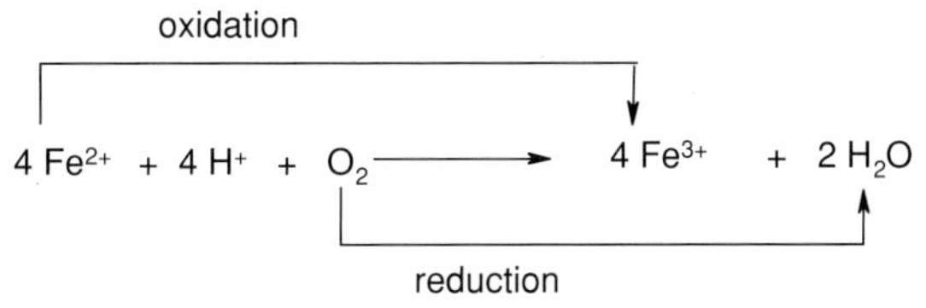

For the oxidation the potential is -0.77 V (as for $Fe^{3+} \longrightarrow Fe^{2+}$ $E^0 = 0.77$ V)

For the reduction the potential is 1.23 V

The overall cell potential is the sum of the two: $E^0_{cell} = -0.77 + 1.23 = 0.36$ V

As E^0_{cell} is positive the reaction is expected to be spontaneous

Figure 1.5b Calculation of cell potentials for a reaction.

Latimer diagrams

For elements with a number of oxidation states available in aqueous solution the **Latimer diagram** is a convenient way of representing its redox properties. An example for the oxidation states of americium is given in Figure 1.5c. Here, we have information on **all** the possible reductions between the species shown. Suppose we require E^0 for Am^{4+}/Am. We cannot simply sum E^0 for each step and take the average. Instead we must calculate ΔG^0 for each step by taking into account the number of electrons transferred, and sum these values to get ΔG^0 for the entire process; then using the above equation calculate E^0. Thus for Am^{4+}/Am^{3+}, $\Delta G^0 = -1F(2.62) = -2.62F$. For Am^{3+}/Am, $\Delta G^0 = -3F(-2.07) = 6.21F$. Hence for Am^{4+}/Am, $\Delta G^0 = -2.62F + 6.21F = 3.59F$. This is a four-electron reduction and therefore $E^0 = -3.59F/4F = -0.90$ V. By this method, in addition to the four reduction potentials given, a further 6 can readily be calculated. The Latimer diagram for americium tells us that each successive reduction from Am(VI) to Am(III) is favourable, but that Am(III) to Am is unfavourable and hence Am^{3+} is the most stable oxidation state. The stability of an oxidation state to **disproportionation** (simultaneous oxidation and reduction) can also be assessed from a Latimer diagram. If the potential involving the reduction is greater than that involving the oxidation then disproportionation will occur. So in the case of Am^{4+}, 2.62 V > 0.82 V and hence the disproportionation of Am^{4+} is spontaneous:

$$2Am^{4+} + 2H_2O \rightleftharpoons AmO_2^+ + Am^{3+} + 4H^+ \quad E^0_{cell} = 2.62 - 0.82 = 1.80\,V$$

The other oxidation states of americium are stable to this type of reaction. It is important to note that whilst the electrochemical data gives information on the final position of this equilibrium, like all thermodynamic methods, it gives no information at all on the **rate** of attainment of this equilibrium. Thus, it is entirely possible to have a large E^0_{cell} and have an unobservably slow reaction!

$$AmO_2^{2+} \xrightarrow{1.60\ V} AmO_2^+ \xrightarrow{0.82\ V} Am^{4+} \xrightarrow{2.62\ V} Am^{3+} \xrightarrow{-2.07\ V} Am$$

Figure 1.5c The Latimer diagram for the oxidation states of americium.

Several other factors affect values of E^0 themselves. The presence of complexing agents can cause enormous changes in redox properties. Cobalt(III) provides the best example of this effect, with the reduction of Co(III) to Co(II) ranging from very favourable to unfavourable depending on the ligand bound to the metal:

$$Co(H_2O)_6{}^{3+}/Co(H_2O)_6{}^{2+} \quad E^0 = 1.90\,V$$
$$Co(NH_3)_6{}^{3+}/Co(NH_3)_6{}^{2+} \quad E^0 = 0.06\,V$$
$$Co(CN)_6{}^{3-}/Co(CN)_5{}^{3-} \quad E^0 = -0.80\,V$$

There is often a considerable dependence of E^0 on the pH of the solutions. Values are generally quoted for standard acidic solution, but in many cases vary considerably at higher pH. For instance, whilst the reduction of O_2 to H_2O has $E^0 = 1.23$ V at pH 0, at pH 14 the value has dropped to 0.41 V, i.e. O_2 is a less powerful oxidant at pH 14.

1.6. Redox Processes (II)

Free energy oxidation state diagrams

Another very useful way of visualising and interpreting redox data for elements with multiple oxidation states is the free energy oxidation state (FROST) diagram. This is a plot of the relative free energy of an element versus its oxidation state. It has some advantages over the Latimer diagram in that it gives an overall visual summary of the relative stability of oxidation states in solution and also gives a global picture of disproportionation, whilst the Latimer diagram only gives the information for adjacent oxidation states. By convention the element itself has zero free energy and its compounds may have higher or lower free energy.

Construction of FROST diagrams

Generally oxidation states of elements are presented as Latimer diagrams and these form the starting point for the construction of the FROST diagram. We plot the free energy change (ΔG^0) for the transformation of each oxidation state to the element ($E^{n+} + ne^- \rightarrow E$) versus the oxidation state. Firstly $\Delta G^0/F$ for each step on the Latimer diagram is calculated (as shown in Section 1.5 for the oxidation states of americium). The example below is for the oxidation states of nitrogen in acidic solution. Here the process for reduction to N_2 is straightforward. To obtain the values for the negative oxidation states **relative to** $N_2 = 0$, we must remember that we are now dealing with an oxidation and hence the sign of E^0 must be reversed as indicated by * in the example below. Once $\Delta G^0/F$ for each step has been calculated the values are summed to give the free energy change in going from the oxidation state to N_2. For example, the point for NO_3^- is calculated as $1.88 + 2.60 + 1.77 = 6.25$ V; that for NH_4^+ as $1.87 - 2.70 = -0.83$ V. These points are plotted against the oxidation state to give a representation of the energy ($\Delta G^0/F$) (Figure 1.6a)

1. E^0		0.94 V		1.30 V		1.77 V		−1.87 V		1.35 V	
	NO_3^-	→	HNO_2	→	N_2O	→	N_2	→	NH_3OH^+	→	NH_4^+
Oxidation state	5		3		1		0		−1		−3
2. Electrons transferred (*n*)	2		2		1		1		2	
3. row 1 × row 2 = nE^0	1.88		2.60		1.77		1.87*		− 2.70*	
ΣnE^0	6.25		4.37		1.77		0		1.87		− 0.83

Interpreting the FROST diagram

There are several points that can be seen at a glance from the FROST diagram. First, the height of an oxidation state is a good indication of its stability. From oxidation state 5 to 0 we are lowering the free energy and thus such reductions are favourable, meaning that species such as nitrate, etc., are good oxidising agents. Similarly, the hydroxylammonium ion, NH_3OH^+ is situated at a peak where both oxidation to N_2 or reduction to the ammonium ion, NH_4^+ are favourable. As the point for NH_3OH^+ lies above the line joining adjacent oxidation states it is thermodynamically unstable with respect to disproportionation. We find a similar situation for all the oxidation states between N_2 and NO_3^-, i.e. they all lie above the line joining N_2 to NO_3^-. On this diagram we can see that N_2O and HNO_2 are thermodynamically unstable to disproportionation to N_2 and NO_3^-, as are other intermediate oxidation states (not shown) such as NO and NO_2.

Just as the high points represent unstable species, those at localised minima represent particularly stable positions. Here, molecular nitrogen is situated at a minimum and is thus expected to be particularly stable, an interpretation that is experimentally verified in the low reactivity of the element. The lowest point on the graph in this case is the lowest oxidation state for

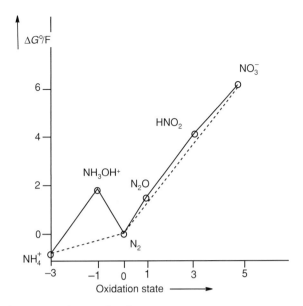

Figure 1.6a The free energy oxidation state diagram for nitrogen.

nitrogen (-3), and NH_4^+ is the most thermodynamically stable species in acidic solution. Throughout this discussion it is vital to remember that this analysis gives only the thermodynamics of the chemistry. Thus, for example, whilst NH_3OH^+ is thermodynamically extremely unstable, its compounds and solutions are **kinetically inert** and hence they can be handled safely and are commercially available.

Figure 1.6b shows the FROST diagram for the oxidation states of americium in acidic solution. Here we can see that the element itself is highly electropositive and that its oxidation to Am^{3+} is extremely favourable. Further oxidation of Am^{3+} is difficult and it represents the most stable oxidation state for this element. Americium(IV) is unstable with respect to disproportionation.

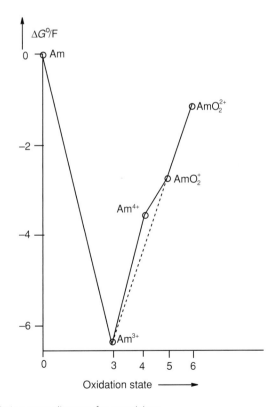

Figure 1.6b The free energy oxidation state diagram for americium.

Molecular Structures and Solid-State Giant Structures

2.1. Covalent Interactions Between Atoms

Overview

The combination (or reaction) of elements with one another leads to materials which are described as **compounds**. Compounds have well-defined (stoichiometric) elemental compositions. Compounds are not simple 'mixtures' of the elements. The structures of compounds may be infinite giant lattice structures or relatively small molecular species, and their properties differ from those of their constituent elements. The valence electrons of the elements are responsible for their reactivity and these interactions are described as **bonding** interactions. For convenience, bonding interactions are commonly referred to as **covalent**, **metallic** or **ionic**. However, it must be appreciated that very few bonding interactions are purely covalent, metallic or ionic and that most have some characteristics of at least two, and possibly all three types. This is illustrated as a triangle in Figure 2.1a. Covalent bonding is described in this section whilst metallic and ionic bonding are described in Sections 2.3 and 2.4, respectively.

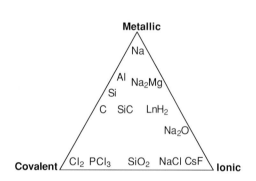

Figure 2.1a Triangle representing the overlap between covalent, ionic and metallic bonding.

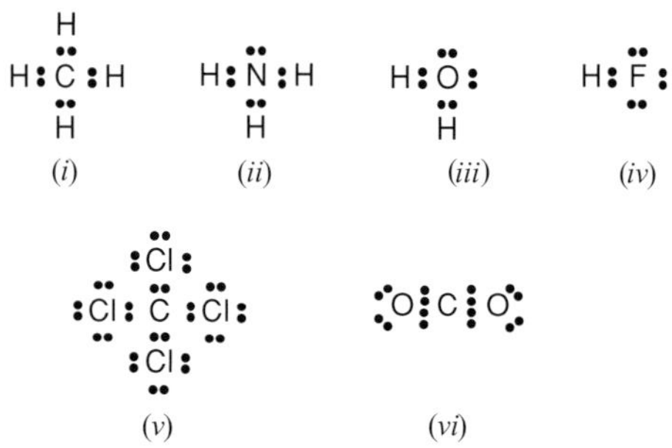

Figure 2.1b Lewis structures of (i) CH_4, (ii) NH_3, (iii) H_2O, (iv) HF, (v) CCl_4 and (vi) CO_2. In (vi) there are $4e^-$ shared between each C and O and this results in a double bond between these elements.

Covalent bonding

The 'inert' gases (Section 4.12) have full outermost electron shells and this electronic configuration is responsible for their stability (i.e. low reactivity). Other elements, particularly main group elements, strive to attain similar full-shell electronic configurations in their compounds. This is commonly achieved by the **sharing of valence electrons** between atoms. The shared electron pair linking two atoms is described as a **bonding pair**, and such a linkage between the two atoms is called a **covalent bond**. Lewis described a simple way of visualising valence electron distribution in elements and molecules with 'dots' representing 'valence electrons', and the 'symbol of the element' to represent the 'nucleus and core electrons'. A basic premise is that, if possible, all electrons should be paired. Pairs of electrons associated solely with one element are termed **lone pairs**. Eight electrons (four pairs) in the valence shell of a main group element is commonly achieved in stable molecules and the element is said to possess a **Lewis octet**. Hydrogen, as a first period element, only needs a share of one more electron to attain the electronic configuration of helium. Examples of Lewis structures for some familiar molecules are shown in Figure 2.1b.

Linear combination of atomic orbitals (LCAO)

A more refined approach to the nature of a bond is given by the LCAO method. Here, atomic orbitals (AOs) on the atoms which are bound together interact to form molecular orbitals (MOs). The number of MOs formed must equal the number of AOs used, and the energies of the MOs are related to the energies of the AOs, the extent of their overlap and their relative phases (signs of the wavefunctions).

16

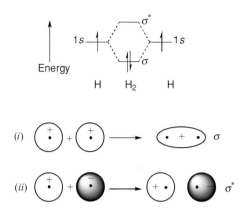

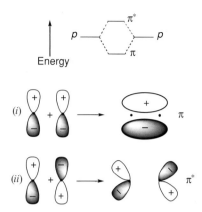

Figure 2.1c The in-phase (i) and out-of-phase (ii) interactions of H $1s$ orbitals in the H_2 molecule.

Figure 2.1d π-bonding interactions between p-orbitals. The 'in-phase' combination (i) results in a low energy (bonding), π, MO. The out-of-phase combination (ii) results in an antibonding, π^*, orbital.

Let us consider the H_2 molecule as a simple example of **homonuclear covalent bonding**. The $1s$ orbitals on each H atom interact to produce two orbitals of different energies. The energies of these orbitals depend upon whether the $1s$ orbitals interact 'in-phase' or 'out-of-phase' (Figure 2.1c). The 'in-phase' combination is of lower energy than the $1s$ orbitals, since this new orbital concentrates electron density between the hydrogen nuclei and holds them together electrostatically. The 'out-of-phase' combination is of higher energy than the $1s$ orbitals, and would, if occupied, have very little electron density between the H atom nuclei. The two hydrogen atoms' two electrons pair up in the lower energy (bonding) MO and the higher energy (antibonding) MO is left vacant. This build up of electron density between the nuclei is characteristic of a σ-bond. The antibonding MO is called a σ^* orbital. 'End-on' interactions of p-orbitals also produce σ and σ^* MOs, whereas a 'side-on' interaction of p-orbitals is weaker but yields π and π^* MOs. In a π MO, electron density is concentrated in planes above and below the axis joining the atoms together (Figure 2.1d).

Heteronuclear covalent bonding follows a similar rationale although the energies of the AOs may be less well matched (electronegative elements have lower energy AOs) and the bonding MO takes on more of the character of the lower energy AO.

Polarity of covalent bonds

In a heteronuclear covalent bond, one atom will invariably be more electronegative than the other and this will lead to unequal sharing of the 'shared' electron pair, with the more electronegative atom having a greater share. This will make the bond **polar** with the more electronegative element at the negative end of the dipole. This has great consequences in intermolecular interactions and on the reactivity of the bond (Section 1.4).

Hybridisation

Hybridisation is a useful concept in that it starts with an element's set of valence AOs and mixes them in such a way as to provide a set of **spatially directed** hybridised orbitals. These hybridised orbitals can be used to form localised MOs (derived from the LCAO approach) to help visualise the σ (and π-bonding) framework of a molecule. A main group element with s, p_x, p_y and p_z AOs in its valence shell might mix them to produce four equivalent sp^3 hybrids (in a tetrahedral arrangement), or three equivalent sp^2 hybrids (in a trigonal planar arrangement) leaving the p_z orbital unmixed, or two equivalent sp hybrids (in a linear arrangement) with the p_x and p_z unmixed (Figure 2.1e). The σ-bonding can make use of the hybrid orbitals, and any π-bonding will use the unhybridised p-orbitals. Other types of hybridisation that might be encountered include sp^2d (square-planar), sp^3d (trigonal bipyramidal) and sp^3d^2 (octahedral).

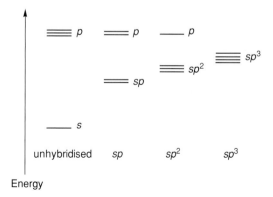

Energy

Figure 2.1e The ns and the three np AOs may mix to yield hybridised AOs as shown. These hybrid orbitals can interact with other atomic orbitals.

17

2.2. Shapes of Molecules

Valence shell electron pair repulsion (VSEPR)

Valence shell electron pair repulsion theory provides a simple model for predicting the shape of molecules/ions that contain a central p-block element (E) surrounded by covalent bonds to adjacent atoms. Lewis structures (Section 2.1) serve as a good basis for understanding VSEPR theory.

Systems involving only single bonds originating from E will be described initially. First, the **number of electrons in the valence shell** of E need to be counted. This is easily accomplished by adding together the number of valence electrons on E (Group number -10, for p-block elements), the number of σ-bonds formed by the element, and any electrons associated with charge (if an ionic species). The electrons in the valence shell of E will be paired-up, as much as possible, and these electron pairs will arrange themselves to be as far away from one another as possible. This is an electrostatic (repulsive) effect since electron pairs are sources of negative charge. The **lowest energy geometry** adopted by n electron pairs is tabulated below and shown diagrammatically in Figure 2.2a. Should there be an unpaired electron in the valence shell of the central atom, then it occupies as much space (i.e. an orbital) as an electron pair.

$n = 2$	linear, 180°
$n = 3$	trigonal planar, 120°
$n = 4$	tetrahedral, 109.5°
$n = 5$	trigonal bipyramidal, 90° and 120°
$n = 6$	octahedral, 90° and 180°

Secondly, the **number of σ-bonds** formed by E is counted and this equals the number of **bonding pairs**. The number of **lone pairs** may be determined by subtracting the number of bonding pairs from the total electron pair count. For systems with $n = 2, 3$ or 4 valence electron pairs, the bonded elements are simply placed in the appropriate number of positions in the expected geometry (as summarised below) where the shape of the molecule is also described (Figure 2.2b). However, the 'space' taken up by a non-bonding pair of electrons is considerably larger than that occupied by a bonding pair (which is constrained by two positively charged nuclei). Since electron pair/electron pair repulsions are dependent upon how far apart the electron pairs are, they decrease in the sequence lone pair/lone pair > lone pair/bonding pair > bonding pair/bonding pair. The result is that lone pairs push bonding pairs closer together than is predicted for the ideal geometry. Representative HEH angles are 109.5° (CH_4), 107° (NH_3) and 105° (H_2O). Electron pair interactions are strongly dependent upon the angle

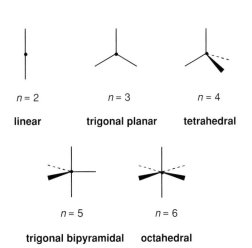

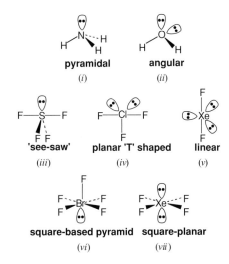

Figure 2.2a The preferred arrangement of n electron pairs around a central atom that minimises electron pair repulsions.

Figure 2.2b Molecular geometries of some molecules illustrating the influence of lone pairs of electrons on the central atom's stereochemistry.

between the electron pairs, since a smaller angle will put the electron pairs closer together (and hence will be disfavoured). In practice, and with all other factors being equal, interactions at 90° or less will dominate the energetics of the systems and larger angles will have much less influence.

Valence electron pairs (n)	Bonding pairs	Lone pairs	Shape	Examples
2	2	0	linear	$BeCl_2$
3	3	0	trigonal planar	BCl_3
4	4	0	tetrahedral	$CH_4, SnCl_4$
4	3	1	pyramidal	NH_3
4	2	2	angular	H_2O, H_2S
5	5	0	trigonal bipyramid	PF_5
5	4	1	'see-saw' geometry	SF_4
5	3	2	planar 'T' shaped	ClF_3
5	2	3	linear	$XeCl_2$
6	6	0	octahedral	$[SiF_6]^{2-}$, SF_6
6	5	1	square-based pyramidal	BrF_5
6	4	2	square-planar	XeF_4, $[ICl_4]^-$

For molecules/ions with **5 valence electron pairs** ($n = 5$), a trigonal bipyramid is the expected geometry and this is observed for PF_5 with five bonding pairs at P. However, SF_4 (with four bonding pairs and one lone pair on S) requires deeper consideration. There are two possible sites for the lone pair to be located in a trigonal bipyrimidal structure (axial or equatorial) and hence SF_4 might exist in one of two isomeric forms (Figure 2.2c). In fact, only one of these structures is observed. Consideration of the relative magnitudes of the electron-pair repulsions as described above indicates which isomer is favoured. If the lone pair is axial there are three lone pair/bonding pair interactions at 90°, whilst if it is equatorial there are only two such 90° interactions. VSEPR predicts that the lone pair will be equatorial, and the 'see-saw' geometry is the observed structure.

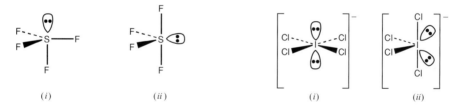

Figure 2.2c The possible isomeric forms of SF_4. Form (ii) is observed due to fewer lone pair–bond pair interactions at 90°.

Figure 2.2d The possible isomers of $[ICl_4]^-$. Form (i) is observed as the lone pair–lone pair interactions are at 180°.

A similar situation arises in $[ICl_4]^-$ which has **six valence electron pairs** arranged octahedrally about the iodine atom (four bonding pairs and two lone pairs). Here the lone pairs might be *cis* or *trans* but a lone pair/lone pair interaction at 90° is very destabilising and the square-planar structure with *trans* lone pairs is the more stable isomer (Figure 2.2d).

In some cases the central element E is involved in multiple bonding (bond order > 1) with $\sigma-$ and π-components, e.g. E=O. The simplest way to account for structures with π-**bonding** is to subtract 1 from the electron count for each π-bond. Thus, for Cl_3PO, the valence shell count about P is 5 (Group number -10) + 4 (σ-bonds) + 0 (charge) -1 (π-bond) = 8 (4 pairs). Thus, to a first approximation the shape adopted by the Cl_3PO molecule is tetrahedral, and the π-component of the double bond has no effect on the overall molecular geometry. However, in reality the π-bond does take up 'extra space' forcing the Cl atoms closer together, in much the same way as a lone pair does in a pyramidal molecule such as NH_3.

2.3. Structure and Bonding in Metals

Structures

Approximately 80% of all elements are classified as metals. **Metallic structures** are best appreciated by considering metal atoms as hard spheres which pack together in regular arrays, forming giant, infinite lattices. In some lattices the atoms are packed together as densely as possible, with minimal 'open space', and these structures are described as **close-packed**. Some metals prefer less densely packed arrangements and such arrangements are described as **non close-packed**.

The number of nearest neighbours a metal has in its structure is defined as its **co-ordination number**. Close-packed structures have a co-ordination number of 12 and non close-packed structures commonly have a co-ordination number of 8. Metallic structures are often described as composed of a series of repeating planes and a metal with a co-ordination number of 12, has 6 nearest neighbours in one plane and 3 nearest neighbours in each plane above and below this 'central' plane. If the top plane and the bottom plane are identical then the metal is an infinite lattice of a series of ...ABAB...planes, and such a structure is described as **hexagonal close-packed (hcp)** (Figure 2.3a). Examples of metals with hcp arrangements are Mg and Zn. An alternative close-packed arrangement is possible where the 'top' and 'bottom' planes are not identical; in such cases the three planes are all different and the resulting infinite lattice is a series of ...ABCABC...planes, and the structure is described as **cubic close-packed (ccp)** or **face-centered cubic (fcc)** (Figure 2.3b). Examples of metals with ccp arrangements are Cu, Au, Pb. Metals with co-ordination number 8 often adopt the non close-packed **body-centered cubic (bcc)** structure (Figure 2.3c), e.g. K and Fe. The **primitive cubic** system is represented by only one example, Po, but this arrangement is often adopted by ions in ionic structures (Figure 2.3d).

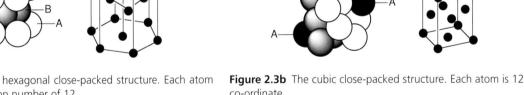

hexagonal close-packed (hcp) CN = 12, repeating layers ...ABABABAB...
cubic close-packed (ccp) CN = 12, repeating layers ...ABCABCABC...

The 'open spaces' found between atoms in close-packed structures are called **interstitial holes**. These holes may be of two types depending upon their geometries: holes surrounded by four atoms are called **tetrahedral holes** and holes surrounded by six atoms are called **octahedral holes**. There are twice as many tetrahedral holes as octahedral holes in close-packed structures (Figure 2.3e). In hcp structures, atoms in layer A sit over tetrahedral holes formed between atoms in lower layers A and B, whilst in ccp structures atoms in layer C sit over octahedral holes formed between atoms in lower layers A and B.

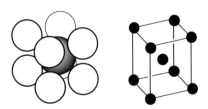

Figure 2.3a The hexagonal close-packed structure. Each atom has a co-ordination number of 12.

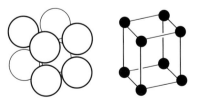

Figure 2.3b The cubic close-packed structure. Each atom is 12 co-ordinate.

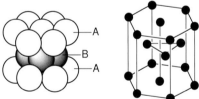

Figure 2.3c The body-centred cube is not a close-packed structure. Each atom is 8 co-ordinate.

Figure 2.3d A primitive cube is a non close-packed structure with a co-ordination number of 6 for each atom.

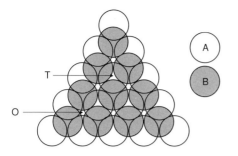

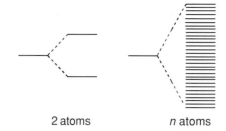

2 atoms   *n* atoms

Figure 2.3e Octahedral (O) and tetrahedral (T) holes are observed between the layers when a close-packed layer B is placed over a close-packed layer A.

Figure 2.3f The interaction of 2 AOs leads to two well-separated MOs. The interaction of *n* AOs in the metal lattice leads to a band of *n* energy levels.

Bonding

A **molecular orbital** (MO) treatment using the LCAO approach gives a useful qualitative picture of bonding between metal atoms in metallic structures. It should be noted that since the co-ordination number of metals in these metallic structures is generally large (8–12), there are insufficient electrons available to form two-electron bond pairs for each metal–metal interaction. **Delocalised MOs**, which extend over the whole lattice, are obtained by the interaction of AOs from the very large number of metal atoms involved (Figure 2.3f). These delocalised MOs take the form of a series of **energy bands**, which often overlap. Maximum bonding interactions occur when the energy bands are half full, and this is reflected in enthalpies of atomisation, and in melting point data, e.g. the Group 6 elements (Cr, Mo, W) are the metals with the highest melting points in each of the three transition-metal series.

Properties

The characteristic physical properties of metals (shiny appearance, high malleability and ductility, high thermal and electrical conductivities) may be attributed to the relatively easy **slippage of the planes** of the metal atoms within the lattice, and to the partially filled electronic **band energy levels**.

2.4. Ionic and Covalent Solid-State Giant Structures

Ionic bonding

In Section 2.1 it was identified that inert gas electron configurations were stable and that elements tend to attain such configurations in their compounds. Any attempt to share electrons between very electronegative elements and very electropositive elements would lead to the formation of very polar bonds. Ultimately, full transfer of the electron from the electropositive element to the electronegative element would occur and this would lead to separated charges and the formation of **positively charged cations** and **negatively charged anions**. The resulting ions are held together in the solid-state as an 'infinite lattice' by electrostatic interactions. A good example is CsF where the Cs^+ cation has the electronic configuration of [Xe], and the F^- anion has the configuration of [Ne].

Simple ionic crystals

Cations and anions arrange themselves in 'infinite' lattices so as to maximise cation–anion attractive forces, minimise cation–cation and anion–anion repulsive forces, and minimise 'empty space'. In many cases this is best achieved by close-packed arrangements of the larger ion (usually the anion) with the smaller ion fitting into some (or all) of the tetrahedral or octahedral holes of the close-packed array. Some important lattices of stoichiometry AB (NaCl, CsCl, ZnS) and AB_2 (CaF_2, TiO_2) are shown in Figures 2.4a–e. The rock salt (NaCl) structure has the Cl^- anions in a face-centered cubic arrangement (cubic close-packed) with Na^+ cations in all the octahedral holes. This is seen clearly by considering the co-ordination of the Na^+ cation in the centre of the cube in Figure 2.4a. The zinc blende (ZnS) structure has the larger S^{2-} ions in a face-centered cubic arrangement with half of the tetrahedral holes occupied by the smaller Zn^{2+} ions. The fluorite (CaF_2) structure has the Ca^{2+} ions in a face-centered cubic arrangement with F^- ions in all the tetrahedral holes. The CsCl and rutile (TiO_2) structures are not close-packed, but the former has an interpenetrating cubic arrangement (e.g. a primitive cube of Cl^- with Cs^+ at the centre), and in the latter the Ti atoms are in a distorted body-centered cubic arrangement.

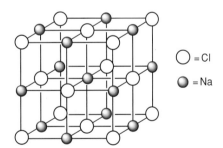

Figure 2.4a The NaCl structure (rock salt).

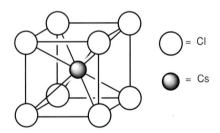

Figure 2.4b The CsCl structure.

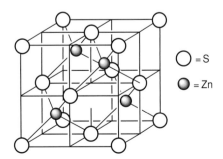

Figure 2.4c The ZnS (zinc blende) structure.

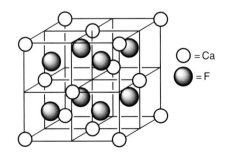

Figure 2.4d The CaF_2 (fluorite) structure.

22

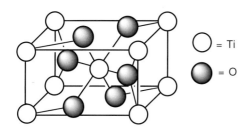

Figure 2.4e The TiO$_2$ (rutile) structure.

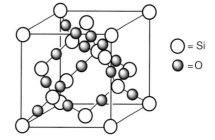

Figure 2.4f The SiO$_2$ (β-cristobalite) structure.

Cation/anion radius ratio rule

If ionic lattice structures are considered as close-packed arrangements of ions of one charge, with the ions of opposite charge fitting into the holes, then there will be an optimum cation/anion size ratio for each given arrangement. More importantly, the size of the holes is limited by their co-ordination number and by the size of the larger ion. The cation/anion radius ratio should give a good indication of a likely stable lattice arrangement for an AB ionic compound as shown below:

$r+/r-$	Co-ordination number of the hole	Structure for AB
0.23–0.41	4	ZnS
0.41–0.73	6	NaCl
>0.73	8	CsCl

Ion polarisability

In Section 2.1 we noted that ionic, covalent and metallic bonding should be considered as extremes in a bonding triangle, and we described the effect of electronegativity on the polarity of a covalent bond. If we now approach the ionic/covalent continuum from the opposite direction, we can consider **polarisation of ions** as the introduction of covalent character into ionic systems.

Partial covalent bonding is observed with small, highly charged cations (high charge/radius ratio), where they are said to **polarise their anionic neighbours** by pulling the anion's electron density towards them. Thus, Be^{2+} is more polarising than Mg^{2+}, and has greater covalent character in its bonds. Larger, more highly charged anions are more polarisable than small anions, since their outer electrons are more easily attracted to the cation as they are further away from the nucleus. Thus, I^{-} is more polarisable than F^{-}. These principles are known as **Fajan's rules**.

Compounds having the most ionic character will therefore be formed by large cations with a small positive charge, paired with small anions carrying a small negative charge, e.g. CsF. Conversely, covalency would be considerable for BeI$_2$ due to polarisation of the large I^{-} anion by the small, highly charged Be^{2+} cation. Covalancy is also considerable in any compound where the formal charge on the cation is $\geq$4.

Covalent giant structures

Not all covalent compounds are (small) molecular species and some are infinite lattice structures. Good examples are (hexagonal) boron nitride (Section 4.4), and graphite and diamond (Section 4.5). Covalent giant structures may be layered (connectivity extending in two dimensions, e.g. graphite) or truly three-dimensional, e.g. diamond. The structure of quartz (SiO$_2$) is shown in Figure 2.4f and it is very similar to that of diamond but has four co-ordinate Si atoms replacing every four co-ordinate C atoms, with two co-ordinate O atoms bridging between each pair of Si atoms. Quartz, like diamond, is a very hard mineral.

SECTION 3

s-Block Elements – Main Group Elements (Groups 1, 2)

3.1. Group 1 Elements – the Alkali Metals (Li, Na, K, Rb, Cs, Fr)

Elements

These elements are soft, low-melting, silvery/white **metals** which conduct heat and electricity well. The elements tarnish rapidly in air and are generally stored under oil. They all have one electron more than an inert gas configuration, and readily lose this electron to form **unipositive ions**. The electropositive nature of these elements increases on descending the group and reactivity of the elements increases correspondingly. The least electropositive element, Li, exhibits covalency in some of its compounds. The elements form very stable ionic compounds and are liberated from them by **electrolysis of their fused chlorides** or hydroxides. The chemistry of Fr, a radioactive element discovered in 1939, is less defined and is not considered here.

Reactivity of the elements

The elements all burn in air, with the more reactive elements (K, Rb, Cs) igniting and readily forming **superoxides** (MO_2). Sodium burns to form a mixture of **peroxide** (Na_2O_2) and **oxide** (Na_2O), whilst Li is less reactive and eventually forms a mixture of **oxide** and **nitride** (Li_3N). The monoxides (M_2O) are basic and react with H_2O to give strongly alkaline metal hydroxide solutions.

$$M_2O(s) + H_2O(l) \rightarrow 2MOH(aq)$$

The reaction of the metallic elements with water leads to formation of the **hydroxide** (MOH) and H_2. With Cs the reaction is explosive, whilst with Li the reaction is slow. However, reactions with acids are more vigorous! Similarly, ammonia reacts to form **amides** (MNH_2), and alcohols e.g. EtOH, ^{t}BuOH, react to form **alkoxides** (MOR).

$$M(s) + H_2O(l) \rightarrow MOH(aq) + \tfrac{1}{2}H_2(g)$$
$$M(s) + NH_3(l) \rightarrow M(NH_2) + \tfrac{1}{2}H_2(g)$$
$$M(s) + EtOH(l) \rightarrow M(OEt) + \tfrac{1}{2}H_2(g)$$

Ionic (saline) **hydrides** (MH) are formed when the metals are heated with H_2. These salts contain the hydride ion (H^-) and give H_2 at the anode when electrolysed in the molten state; they are hydrolysed by H_2O to MOH and H_2.

Compounds

The **halides** (MX) are crystalline, ionic, high-melting solids which are conveniently prepared by the action of aqueous hydrohalic acid (HX) on the metal carbonate (M_2CO_3) or hydroxide. Lithium chloride is deliquescent and partly hydrolysed in aqueous solution due to the strongly polarising Li^+ cation.

$$LiCl + H_2O \rightleftharpoons LiOH + HCl.$$

The **nitrates** (MNO_3) may be prepared by a similar route using HNO_3 and an alkali metal carbonate or hydroxide. They decompose on heating to afford the nitrite.

$$MNO_3 \rightarrow MNO_2 + \tfrac{1}{2}O_2$$

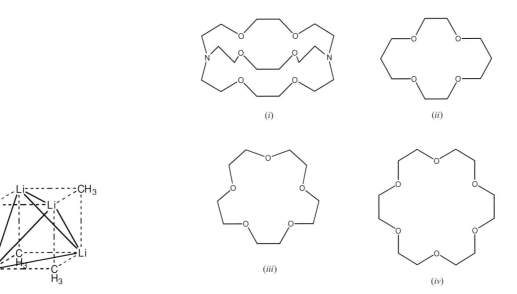

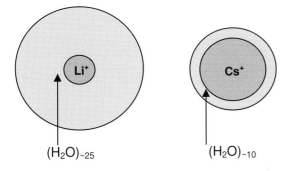

Figure 3.1a Schematic representation of the solid-state structure of Me$_4$Li$_4$. The four Li atoms are arranged in a tetrahedron with Me groups bridging each of the four faces, giving a distorted cube-like structure for the Li$_4$C$_4$ core.

Figure 3.1b Schematic drawings (i) of a bicyclic cryptate, (ii) 14-crown-4, (iii) 15-crown-5, (iv) 18-crown-6. The 'hole sizes' in the latter three are comparable in size to those of the Li$^+$, Na$^+$ and K$^+$ ions, respectively, and form strong complexes with these ions.

Figure 3.1c Schematic drawing showing the relative size of the secondary co-ordination shells of two solvated alkali metal cations, Li$^+$(aq) and Cs$^+$(aq).

The alkali metal **carbonates** (M$_2$CO$_3$, M = Na, K, Rb, Cs) are the only readily soluble metal carbonates, and unlike other metal carbonates they generally do not decompose on moderate heating to the oxide. However, Li$_2$CO$_3$ is only sparingly soluble, and does decompose on heating.

Lithium will react with aryl or alkyl halides in dry ether or THF to afford **organometallic** species with largely covalent Li–C bonds; these species are usually not simple monomers but aggregates e.g. Me$_4$Li$_4$, (Figure 3.1a). Other alkali metals form organometallic compounds that are appreciably ionic.

Crystalline species such as [Na(cryptate)]$^+$Na$^-$ containing the **sodide** anion are isolatable at low temperature. A drawing of a cryptate ligand is shown in Figure 3.1b.

Co-ordination complexes

In aqueous solution Li$^+$, Na$^+$ and K$^+$ have a **primary co-ordination** number of 4, and Rb$^+$ and Cs$^+$ probably 6. All ions have additional layers of H$_2$O molecules bound to the hydrated metal ions in so-called **secondary solvation layers**. The charge density of the hydrated M$^+$ cations have a great influence on the secondary solvation layers, with hydration numbers and the effective radius of the M$^+$(aq) ions following the order Li$^+$(aq) > Na$^+$(aq) > K$^+$(aq) > Rb$^+$(aq) ∼ Cs$^+$(aq) (Figure 3.1c).

Ethers (e.g. THF), polyethers (glyme, diglyme) and cyclic polyethers (crown ethers) (Figure 3.1b) and other **polydentate donors with hard O donor atoms** are ideally suited to complex the hard alkali metal cations. Complexation in this manner enhances the solubility of the alkali metal salts in non-aqueous solvents.

Flame test colours

Flames display **characteristic colours** when alkali metal salts e.g. chlorides, are placed within them. This is referred to as a **flame test**. The colours are due to thermal excitation of electrons within the alkali metal atom/ion. This effect is used in emission flame photometry and atomic absorption spectroscopy for quantitative measurements.

Li – crimson; Na – yellow; K – lilac; Rb – violet-red; Cs – blue

27

3.2. Group 2 Elements – the Alkaline Earths (Be, Mg, Ca, Sr, Ba, Ra)

Elements

These elements are all soft, silvery/white **metals**, with melting points higher than those of their neighbouring Group 1 elements. They have two electrons in their outer shell and the **+2 oxidation state** dominates their chemistry, with ionic compounds the norm for the heavier, more reactive, more electropositive metals (Mg, Ca, Sr, Ba). Their reactivity increases down the group. The less reactive, lighter metals (Be, Mg) also show some covalent properties as a result of the highly polarising nature of these smaller M^{2+} ions. All the elements may be extracted from their **fused chlorides by electrolysis** and they are also obtainable by **high temperature chemical reduction** of their oxides. Calcium is the fifth most abundant element in the Earth's crust and is often found as limestone ($CaCO_3$). Radium is radioactive.

Reactivity of the elements

The alkaline earths will combust in air to yield mixtures of the **oxides** (MO) and **nitrides** (M_3N_2). The heavier alkaline earths (Ca, Sr, Ba) burn in O_2 to produce **peroxides** (MO_2), in addition to MO. However, a better method of preparation of their oxides is by thermal decomposition of their carbonates (MCO_3).

$$MCO_3 \rightarrow MO + CO_2$$

Whereas Be is unreactive towards H_2O, the heavier alkaline earths (Ca, Sr, Ba) react readily to afford their **hydroxides**, $M(OH)_2$, and H_2. Magnesium will only react with steam at red-heat.

$$M + 2H_2O \rightarrow M(OH)_2 + H_2$$

The elements react readily with acids to yield $M^{2+}(aq)$ ions and H_2. The reaction of Ba with H_2SO_4 is slow because of formation of the insoluble sulphate, $BaSO_4$. Beryllium displays its **amphoteric** nature by also dissolving in acids to form Be^{2+} salts, and in alkali to form **beryllates**, $[Be(OH)_4]^{2-}$.

With the exception of Be, the alkaline earths react when heated in H_2 to form the **dihydrides**, MH_2. The hydrides of the heavier alkaline earths (Ca, Sr, Ba) are ionic and hydridic, but MgH_2 is partially covalent. Beryllium dihydride exists but is less stable, and is best prepared by thermolysis of $Be(^tBu)_2$; it is a covalent polymer (Figure 3.2a).

Compounds

The **oxides**, MO, of the heavier alkaline earths (Mg, Ca, Sr, Ba) are basic. They are obtained by thermal decomposition of carbonates, hydroxides, nitrates or sulphates.

The **hydroxides**, $M(OH)_2$, formed by the reaction of oxides with H_2O, have limited solubility. The solubility of the hydroxides increases upon descending the group, but they may all be precipitated out of solution by addition of alkali, e.g. NaOH(aq), to $M^{2+}(aq)$ ions. BeO is amphoteric and will dissolve in alkali and acid.

The **nitrates**, $M(NO_3)_2$, prepared by the action of HNO_3 on carbonates, hydroxides or the oxides, all decompose on heating with formation of the oxide.

$$2M(NO_3)_2 \rightarrow 2MO + O_2 + 4NO_2$$

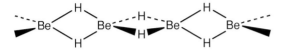

Figure 3.2a Part of the polymeric chain found in solid BeH_2.

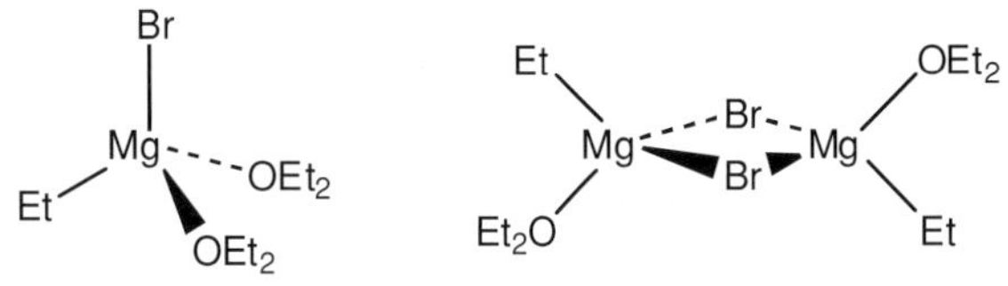

Figure 3.2b The structures of (i) $[EtMgBr(OEt_2)]_2$ as observed in the solid-state. In dilute solution structure (i) persists but at higher concentrations dimers (structure ii) and even more associated species are abundant. The association is likely to occur via halide bridges.

Carbonates (MCO_3) may be precipitated out of solution by the addition of Na_2CO_3(aq) to solutions containing M^{2+}(aq) ions.

The **halides** of the heavier alkaline earths (Mg, Ca, Sr, Ba) are essentially crystalline ionic solids containing M^{2+} and X^- ions, which hydrate in aqueous solution. Beryllium chloride is a covalent polymer, which is readily hydrolysed to give acidic solutions.

Beryllium and magnesium form **organometallic** species with polar covalent M−C bonds. The most important are called **Grignard reagents**, **RMgX**, obtained by the interaction of Mg with alkyl or aryl halides in THF or ether solvents. The Mg centre is generally tetrahedral e.g. $RMgX(OEt_2)_2$, (Figure 3.2b). They have considerable use in organic synthesis as reagents for the formation of C−C bonds. The organometallic chemistry of Ca, Sr and Ba is less well-developed.

Co-ordination complexes

In aqueous solution the alkaline earth metals exist as **aqua ions**, $[M(H_2O)_n]^{2+}$, with n commonly 4 (Be), 6 (Mg, Ca) or 6 and higher (Sr, Ba). The 'hard' alkaline earth metal ions prefer hard O or N donor atoms, but stability constants generally decrease on descending Group 2 due to the larger sizes of the heavier metal ions, leading to weaker electrostatic interactions with the ligands. EDTA readily forms complexes with Mg^{2+} and Ca^{2+} in aqueous solution, and this is made use of in their titrimetric analysis (Figure 3.2c).

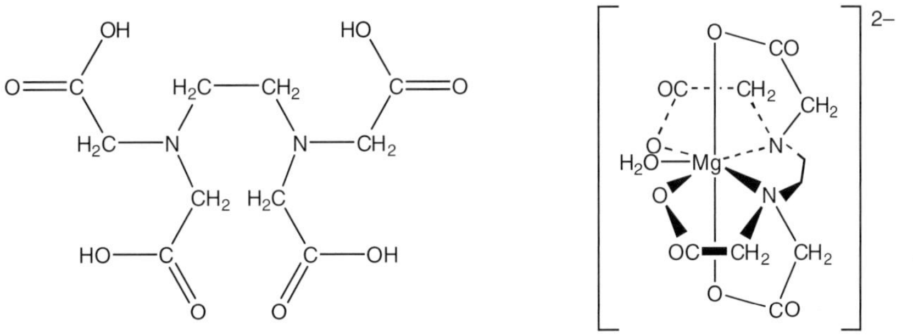

Figure 3.2c Schematic diagrams of (i) H_4EDTA and (ii) $[Mg(EDTA)(H_2O)]^{2-}$. The co-ordination number of Mg^{2+} is 7. For the larger Ca^{2+} ion, the co-ordination number is 8 and there is an additional H_2O molecule in the first co-ordination sphere. Formation constants (β) for complexes of Mg^{2+} and Ca^{2+} with $EDTA^{4-}$ are $10^{8.79}$ and $10^{10.69}$, respectively.

Flame test colours

Flames display **characteristic colours** when some alkaline earth metal salts, e.g. chlorides, are placed within them. The colours are due to thermal excitation of electrons within the alkaline earth atom/ion.

Ca – brick red; Sr – scarlet; Ba – yellowish green

3.3. Some Industrial Processes Involving *s*-Block Elements

Production and uses of salt

Salt (NaCl) is primarily obtained from solar evaporation of seawater (brine), or mined as **rock salt** from localised deposits. The main use of salt is in the manufacture of caustic soda (chloralkali industry). 'Gritting' of roads for snow/ice clearance and the manufacture of soda ash (Solvay process) are also important uses.

Manufacture of caustic soda and the chloralkali industry

Caustic soda (NaOH) is made by electrolysing a strong brine solution in a chloralkali cell with Cl_2 and H_2 as useful by-products. Chlorine is liberated at the anode and OH^- and H_2 are obtained at the cathode. The diaphragm cell (Figure 3.3a) and the mercury cell are the two principal types of chloralkali cells in current use, with the former favoured for both economic and environmental reasons.

Anode reaction	Cathode reactions
$Cl^-(aq) \rightarrow \frac{1}{2}Cl_2(g) + 1e^-$	$Na^+(aq) + 1e^- \rightarrow [Na]$
	$[Na] + H_2O \rightarrow Na^+(aq) + OH^-(aq) + \frac{1}{2}H_2(g)$
	Overall: $H_2O + 1e^- \rightarrow OH^-(aq) + \frac{1}{2}H_2(g)$

Extraction of Na

Sodium is obtained from fused NaCl by **electrolysis** in a **Downs cell** (Figure 3.3b). To reduce energy costs $CaCl_2$ is added to the NaCl (3:2 mixture) in order to reduce the fusion temperature to ~580°C. Pure NaCl melts at >800°C. Calcium is also produced at the cathode but it is returned to the melt. Large-scale production of Cl_2 is also carried out using the Downs process.

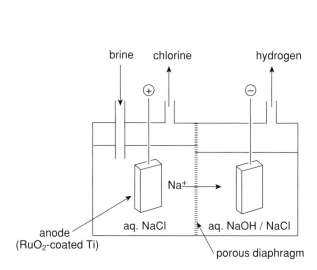

Figure 3.3a A diaphragm cell as used in the chloralkali industry for the production of Cl_2 and caustic soda (NaOH).

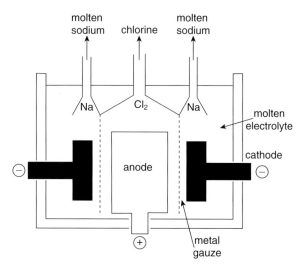

Figure 3.3b A Downs cell for the production of Na from NaCl. Recombination of electrolysis products (Na and Cl_2) is avoided by keeping them separate.

Soda ash

Until recently, all **soda ash** (Na_2CO_3) was made via the **Solvay process**, which consumes brine, limestone and energy, and recycles NH_3. The Solvay process is a multi-step process with the overall reaction being

$$2NaCl + CaCO_3 \rightarrow Na_2CO_3 + CaCl_2$$

Direct mining of 'trona', $Na_2CO_3 \cdot NaHCO_3 \cdot 2H_2O$, and its thermal decomposition to soda ash is now generally a more economical process.

Sodium carbonate is mildly alkaline, and in its hydrated form, $Na_2CO_3 \cdot 10H_2O$, is familiar in **washing soda** as large, colourless crystals. However, its main use is in the glass industry where it, along with silica and lime, is a major constituent of soda-lime glass.

Lime

White chalky sedimentary rocks (**limestone**) contain large amounts of $CaCO_3$ and are an important source of **lime** (CaO). The lime is obtained from limestone in lime kilns by a high temperature ($\sim 1000°C$) endothermic reaction which involves elimination of CO_2. The CO_2 is usually lost to the atmosphere (greenhouse gas!), but may be used as a reactant for an on-site Solvay process. Freshly prepared lime is often referred to as **quicklime**.

$$CaCO_3(s) \rightarrow CaO(s) + CO_2(g)$$

Calcium oxide is a basic metal oxide that reacts very exothermically with H_2O to produce **slaked lime**, $Ca(OH)_2$, which is moderately soluble in H_2O. Its solutions are strongly alkaline.

$$CaO(s) + H_2O(l) \rightarrow Ca(OH)_2(s)$$
$$Ca(OH)_2(s) + excess\ H_2O \rightarrow Ca^{2+}(aq) + 2OH^-(aq)$$

Limestone is quarried on an enormous scale and is widely used as a building material and as a source of lime and quicklime. The principal uses of lime are in cement making, the steel industry, paper and pulp making industries and the glass industry.

Extraction of Mg

Magnesium is the third most abundant element in **seawater** and can be extracted from it (**Dow process**) by chemical treatment followed by electrolysis. Slaked lime is used to precipitate out $Mg(OH)_2$ which is then neutralised by HCl with formation, after drying and heating, of anhydrous $MgCl_2$. Electrolysis of the fused anhydrous $MgCl_2$ at $750°C$ affords Mg at the cathode and Cl_2 at the anode.

$$Mg^{2+}(aq) + 2OH^-(aq) \rightarrow Mg(OH)_2(s)$$
$$Mg(OH)_2(s) + 2HCl(aq) \rightarrow MgCl_2.xH_2O(s) \rightarrow MgCl_2(s)$$
$$\underline{anode}: 2Cl^-(l) \rightarrow Cl_2(g) + 2e^- \quad \underline{cathode}: Mg^{2+}(l) + 2e^- \rightarrow Mg(l)$$

Magnesium may also be obtained from the mineral **dolomite $MgCO_3 \cdot CaCO_3$** by the **silicothermal process** involving thermal conversion to the mixed oxide followed by chemical reduction with a ferrosilicon alloy under reduced pressure at $1150°C$. The Mg is obtained by distillation.

$$CaCO_3 \cdot MgCO_3 \rightarrow CaO \cdot MgO + 2CO_2$$
$$2CaO \cdot MgO + FeSi \rightarrow 2Mg + Ca_2SiO_4 + Fe$$

p-Block Elements – Main Group Elements (Groups 13–18)

4.1. General Overview of *p*-Block Chemistry (I)

Electronic configurations

The *p*-block is probably the most diverse in the Periodic Table in terms of both the physical and chemical properties of its elements. Chemical **reactivity** varies from the most reactive element, fluorine, to the least reactive, helium.

Electronic configurations are as indicated below, with the core *d*- and *f*-subshells shown, as they have a considerable bearing on the chemistry of the *p*-block.

$ns^2 np^x$	$(n = 2, 3)$
$[(n-1)d^{10}] ns^2 np^x$	$(n = 4, 5)$
$[(n-2)f^{14}(n-1)d^{10}] ns^2 np^x$	$(n = 6)$

Ionisation energies

Whilst ionisation energies do not fully explain the chemistry of any element, they can give an insight into some aspects of observed reactivity, see Section 1.4.

The trend going **down** any particular group is more complex than that found for the *s*-block metals. Whilst the general trend is the expected decrease, there are significant deviations. These are particularly pronounced when ionisation from *s*-orbitals is involved, for instance the fourth ionisation energy, I(4) for Bi is higher than that for antimony, whilst the third ionisation energy, I(3), for Pb and Ge are higher than those for Sn and Si, respectively, see Figure 4.1a. These trends can be explained by an **increased effective nuclear charge**, experienced as a result of the introduction of the first *d*- and *f*-block elements in periods 4 and 6, respectively, where an extra 10 and 14 elements are introduced prior to the *p*-block. The corresponding increase in nuclear charge is not fully offset by the filled 3*d*- and 4*f*-orbitals which, as they are the first *d*- and *f*-orbitals, are relatively poor at shielding the valence *s*- and *p*-electrons. This increase is particularly strongly felt by the *s*-electrons as these have non-zero electron density at the nucleus. In addition **relativistic effects** are thought to be significant in period 6. Here electrons orbiting close to a highly charged nucleus will be moving at a velocity comparable to that of light. Objects travelling at such speeds have an increased mass according to the equation:

$$M_V = \frac{M_0}{\sqrt{(1 - V^2/c^2)}}$$

where M_V and M_0 are the masses at velocity V and the rest mass, respectively, and c is the velocity of light. As V approaches c, M_V becomes very large. The chemical significance of this is that the binding energy of the electron is directly proportional to its mass and hence related to its velocity.

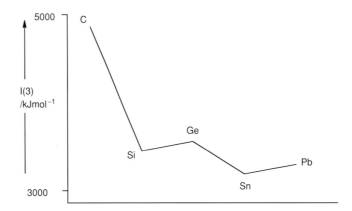

Figure 4.1a The trend in 3rd ionisation energy, I(3), in Group 14 elements.

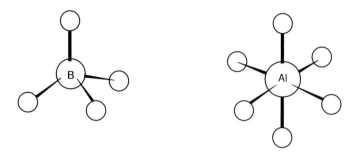

Figure 4.1b   The structures of the BF_4^- and AlF_6^{3-} ions, showing the expansion of the octet around aluminium.

As a result of the higher ionisation energies the $4s$ and $6s$ electrons are reluctant to become involved in bonding, and the high oxidation states of period 4 and 6 *p*-block elements are considerably more oxidising then their period 3 and 5 counterparts. This can be seen in the stability of compounds and also in their **redox properties** (see Sections 1.5 and 1.6). The weak bonds formed with other elements will also be a factor in the lower stability of many of the compounds of the heavier elements. Below are some illustrative examples of stability and redox behaviour from Groups 15 and 17.

$AsCl_5$ is unstable above $-30°C$ and $BiCl_5$ has not been observed.
PCl_5 and $SbCl_5$ are stable at normal temperatures.

Bismuth(V) is a powerful oxidant – BiO_3^-/Bi^{3+} $E° = 2.0$ V
Antimony(V) is a much weaker oxidant – Sb_2O_5/Sb_4O_6 $E° = 0.70$ V

Perbromate is a more powerful oxidant than perchlorate
BrO_4^-/BrO_3^- $E° = 1.85$ V, ClO_4^-/ClO_3^- $E° = 1.20$ V

The trends in ionisation energies also help to explain the transition from metallic elements to non-metals. Where the ionisation energies are relatively low it is fairly easy for electrons to enter the conduction band; thus metallic character decreases as ionisation energies rise moving across a period, whilst the general decrease on descending a group increases the metallic character.

Bonding

Second period elements tend to form compounds that have eight electrons in their valence shell (octet rule) and thus four bonds to other elements, e.g. CO_2, CH_4, BF_4^-, or three bonds plus one lone pair in NH_3, two bonds plus two lone pairs in H_2O and so on. The heavier elements are capable of undergoing a so-called **expansion of the octet** to give compounds with higher co-ordination numbers, such as PF_5, SF_6, IF_7, XeF_6 and many others, see Figure 4.1b. The reasons for these properties are not yet fully agreed upon. Some explanations involve the use of relatively low energy *d*-orbitals, the large size of the atoms, or **multicentre bonding** to accommodate the extra bonded atoms. With the latter approach the phosphorus atom in PF_5 uses its four valence orbitals to bond to the five fluorine atoms, and there is no need to invoke high-energy *d*-orbitals on the phosphorus (Figure 4.1c).

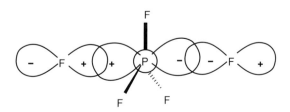

Figure 4.1c   The proposed multicentre bonding between the axial fluorines and phosphorus in PF_5.

4.2. General Overview of *p*-Block Chemistry (II)

Multiple bonds

There are significant differences in the properties of period 2 elements, B to F, compared with the heavier members of their groups. This is seen in the structures of the elements themselves, and the properties of their compounds. For example, the tendency to form **strong double bonds** is much more pronounced for oxygen than for sulfur, selenium or tellurium, a feature which is common to all the *p*-block elements. Thus oxygen exists as a diatomic doubly bonded molecule, O_2, whilst sulfur exists as a singly bonded cyclic molecule, S_8.

The reason lies in the relative stability of the single and double bonds. Oxygen forms strong double bonds due to the **good $2p/2p$ π-overlap** made possible by the short O−O single bond distance. Whilst the same type of π-bonding is possible for S, Se and Te, the longer S−S, Se−Se and Te−Te single bonds mean that the larger and more diffuse 3*p*-, 4*p*- and 5*p*-orbitals cannot approach closely enough to give a significant overlap and hence give only weak double bonds, as illustrated in Figure 4.2a. This is also reflected in the structures of other *p*-block elements such as $N_2(g)$, with a triple bond, compared to $P_4(s)$. Some structures of singly bonded allotropes are shown in Figure 4.2b.

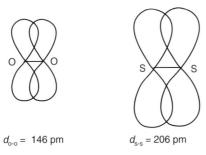

$d_{\text{o-o}}$ = 146 pm $\qquad$ $d_{\text{s-s}}$ = 206 pm

Figure 4.2a Longer bond distances lead to less effective *p*-orbital overlap (as a percentage of the volume of the orbital) for period 3 and heavier elements and hence weaker double bonds.

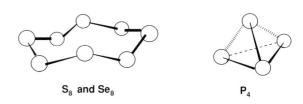

S_8 and Se_8 $\qquad\qquad$ P_4

Figure 4.2b The structures of S_8 and Se_8 are cyclic; Te forms a polymeric chain with single Te−Te bonds. The structure of white phosphorus is tetrameric, P_4; As_4 exists in the gas phase. The **most** stable structures for P, As, Sb and Bi are all singly bonded polymeric networks.

Multiple bonds to other elements tend to be strong between two period 2 elements, for example C=C, C=O, C=N, N=O, etc., and whilst multiple bonds between period 2 and the elements in subsequent periods can be formulated, their exact nature is often not as simple to describe. For instance, phosphine oxides R_3PO are often assumed to have a formal double bond between P and O, i.e. P=O, but many of their properties imply that this is better formulated with a charge separation as $P^+ - O^-$.

In many cases, whether compounds adopt doubly or singly bonded structures is a delicate balance depending on the relative strengths of the appropriate single and double bonds. Thus, whilst oxygen forms double bonds to carbon to give CO_2, a discrete molecule, the structure of SiO_2 is a giant polymeric structure with no double bonds between Si and O. Here it is more energetically favourable to form 4 × Si−O rather than 2 × Si=O; see Section 4.6.

Bond enthalpies are a measure of the heat **required** to break a bond and can be used to rationalise why one structure is preferred over another. For example P_4O_{10} is the stable form of phosphorus(V) oxide. This can be understood by considering the enthalpy change for the hypothetical reaction $P_4O_{10} \rightarrow 2P_2O_5$ (Figure 4.2c). To break all the bonds in P_4O_{10} requires $4D_{P=O} + 12D_{P-O}$ (breaking four double and twelve single bonds) where D is the bond enthalpy. On re-forming 2 moles of P_2O_5 the heat equivalent to $2[4D_{P=O} + 2D_{P-O}]$ would be released. Thus, the overall enthalpy change would be:

$$\Delta H = 4D_{P=O} + 12D_{P-O} - 2[4D_{P=O} + 2D_{P-O}] = -4D_{P=O} + 8D_{P-O}$$

The values for bond enthalpies are $D_{P=O} = 599$ kJmol^{-1} and $D_{P-O} = 340$ kJmol^{-1} and hence $\Delta H = +324$ kJmol^{-1}. The reaction would be endothermic and thus P_4O_{10} would not be expected to convert to P_2O_5. The structure of nitrogen(V) oxide, however, is N_2O_5 as a consequence of the stronger N=O bond.

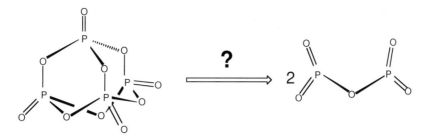

Figure 4.2c The relative stabilities of P_4O_{10} versus P_2O_5 can be assessed using bond enthalpies.

Molecular geometries

Whilst the compounds of *p*-block elements display a wide variety of geometries, two common ones represent the majority of structures. The tetrahedral and octahedral arrangements are particularly favoured, as will be seen in the following sections. Even where structures seem at first sight to be rather complex, they can often be considered to be made up from simpler tetrahedral and octahedral arrangements, and some examples are shown in Figure 4.2d. For instance the solid-state structure of $SnCl_2$ has tin bonded to three chlorines with one lone pair giving a geometry based on a tetrahedron. There are many examples of EX_3 species where E bears a lone pair, for instance ER_3 (E = Group 15 element, R = H, halogen, organic, alkoxy, etc; E = Group 17 element, X = O).

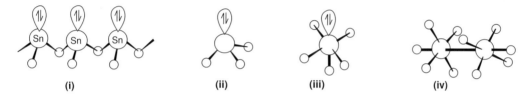

| (i) | (ii) | (iii) | (iv) |

Figure 4.2d Some structures based on tetrahedral and octahedral distributions of electron pairs. (i) $SnCl_2$, (ii) NH_3, (iii) BrF_5, (iv) S_2F_{10} and Se_2F_{10}.

4.3. Group 13 Elements (B, Al, Ga, In, Tl)

Overview

The principal oxidation state for the Group is +3 and the halides and oxides have the expected formulae, EX_3 and E_2O_3. Thallium(III) is a strong oxidant and Tl(I) has a significant chemistry.

Halides

BX_3 derivatives are all **volatile monomers** whilst the fluorides of the heavier elements (Al, Ga, In, Tl) have ionic lattice-like structures and their chlorides, bromides and iodides are dimeric in the solid and liquid states. Two typical structures are shown in Figure 4.3a. The reason for these differences probably lies in the size of the atoms and the nuclear repulsions at close proximity that the short bond distances in BX_3 would require for the dimer. For Al and subsequent elements the longer $E-X$ distances reduce both steric crowding and nuclear repulsion. In the dimers there are four bonds to the central atom and hence all four valence orbitals are used in bonding. In BX_3 there is a tendency to utilise the 'vacant' p-orbital where possible. In BF_3 the $B-F$ bond distance is significantly shorter than expected due to $2p/2p$ π-bonding, as shown in Figure 4.3b.

As the halides formally possess a vacant, or potentially vacant, p-orbital they are **strong Lewis acids** and form complexes with a variety of neutral and anionic donors. The Lewis acid strength increases along the series $BF_3 < BCl_3 < BBr_3 < BI_3$ due to the decreasing π-bonding between boron and the halogen (see Section 4.2). If the donor contains an acidic hydrogen, **solvolysis/hydrolysis** of the $E-X$ bond may occur with loss of HX as indicated below.

$$EX_3 + 3ROH \rightarrow E(OR)_3 + 3HX$$
$$EX_3 + H_2O \rightarrow E(OH)_3 + 3HX$$

Compounds such as InX_2 and TlI_3 illustrate the increasing **oxidising ability of the +3 oxidation state.** Compound InX_2 exists as $In^+[InX_4]^-$, containing In(I) and In(III), whilst TlI_3 is actually a thallium(I) compound containing the triiodide ion, $Tl^+[I_3]^-$. The reduction potential for $Tl^{3+}/Tl^+ = 1.25$ V which confirms the strongly oxidising nature of Tl(III). The oxidising ability is strongly dependent on the co-ordination environment (as with other metals, see Section 1.5), and whilst Tl^{3+} should readily oxidise iodide, the complex ion $[TlI_4]^-$ is stable even though TlI_3 is unknown. Thallium(I) compounds have similar properties to those of silver(I). The halides are photosensitive and insoluble in water, apart from the fluoride which is very soluble.

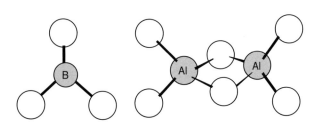

Figure 4.3a Boron trichloride is monomeric. The longer Al−Cl distances lead to lower steric crowding around Al and reduce the repulsions between the Al nuclei in Al_2Cl_6.

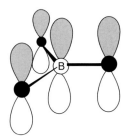

Figure 4.3b The π-bonding in BF_3 utilises the vacant p-orbital on boron; all 4 valence orbitals on boron are used in bonding to fluorine.

Borates and aqua ions

In its aqueous chemistry boron behaves as a non metal. Orthoboric acid, $B(OH)_3$ acts as a weak acid by acting as an **OH⁻ acceptor** rather than as a proton donor:

$$B(OH)_3 + 2H_2O \rightarrow B(OH)_4^- + H_3O^+ \quad pK_a = 9$$

A large number of anionic compounds exist in the solid state, some of which are shown in Figure 4.3c. Common structural features are the presence of **tetrahedral** BO_4 and **trigonal planar** BO_3 units. The most commercially important is the tetraborate anion $[B_4O_5(OH)_4]^{2-}$ as found in borax. Because of the high charge to size ratio, there are no compounds containing simple B^{3+}(aq) ions. This contrasts with the remainder of the Group, where $[M(OH_2)_6]^{3+}$ ions are significant.

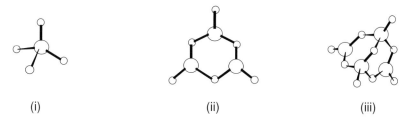

(i)   (ii)   (iii)

Figure 4.3c The structures of some borate anions showing the geometry around the boron atoms. (i) $[B(OH)_4]^-$ and $[BO_4]^-$, (ii) $[B_3O_6]^{3-}$ and (iii) $[B_4O_5(OH)_4]^{2-}$ (H-atoms omitted for clarity).

Co-ordination chemistry

The hexaqua ions tend to undergo acid hydrolysis and substitution reactions:

$$[M(OH_2)_6]^{3+} + H_2O \rightarrow [M(OH_2)_5(OH)]^{2+} + H_3O^+$$
$$[M(OH_2)_6]^{3+} + \text{excess } Cl^- \rightarrow [MCl_4]^- \quad (M = Al, Ga)$$
$$[M(OH_2)_6]^{3+} + \text{excess } Cl^- \rightarrow [MCl_5]^{2-} + [MCl_6]^{3-} \quad (M = In, Tl)$$

The co-ordination numbers of the chloride complexes are dependent on the ionic radii of the metal. There is an extensive **co-ordination chemistry** of metals with organic ligands. Two important examples are complexes with pentane-2,4-dione, $(CH_3C(O)CH_2C(O)CH_3$, acetylacetone, acac) and 8-hydroyquinoline, both of which lose an acidic hydrogen on co-ordination to give neutral octahedral complexes, shown in Figure 4.3d.

$$Al^{3+} + 3HL \rightleftharpoons [AlL_3] + 3H^+$$

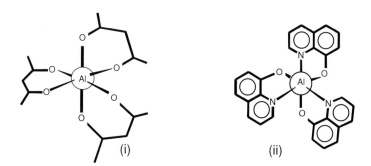

(i)   (ii)

Figure 4.3d The structures of (i) $[Al(acac)_3]$ and (ii) $[Al(8\text{-hydroxyquinolate})_3]$. $[Al(acac)_3]$ is appreciably volatile and $[Al(8\text{-hydroxyquinolate})_3]$ is used in the gravimetric analysis of aluminium.

4.4. Borazine and Boron Hydrides

Borazine, $B_3N_3H_6$

As boron has one electron fewer than carbon, and nitrogen one more, BN compounds are thus **isoelectronic** with corresponding carbon compounds. Thus, the structure of hexagonal boron nitride (BN) is very similar to that of graphite, with layered hexagonal sheets; and the cubic form of BN is isostructural with diamond. Analogues of simple aromatic systems are well known, for example $B_3N_3H_6$ (borazine) is isoelectronic with benzene, C_6H_6, has a similar planar structure (Figure 4.4a) and some similarities in chemical reactivity. It can be prepared by reaction of BCl_3 with ammonium chloride followed by reduction with $LiAlH_4$:

$$3BCl_3 \quad + \quad 3NH_4Cl \quad \xrightarrow{\text{heat}} \quad (BClNH)_3 \quad \xrightarrow{\text{LiAlH}_4} \quad (BHNH)_3$$

B–N 144 pm, N–H 102 pm, B–H 120 pm   C–C 142 pm, C–H 108 pm

Figure 4.4a The structure of borazine compared with that of benzene. The double bond can be thought of as arising from the donation of the N lone pair into the empty *p*-orbital on the boron.

Unlike benzene the bonds are significantly polar and borazine is thus more reactive. It readily undergoes addition reactions with water and HCl, Figure 4.4b, and is hydrolysed by excess water to give $B(OH)_3$ and NH_3.

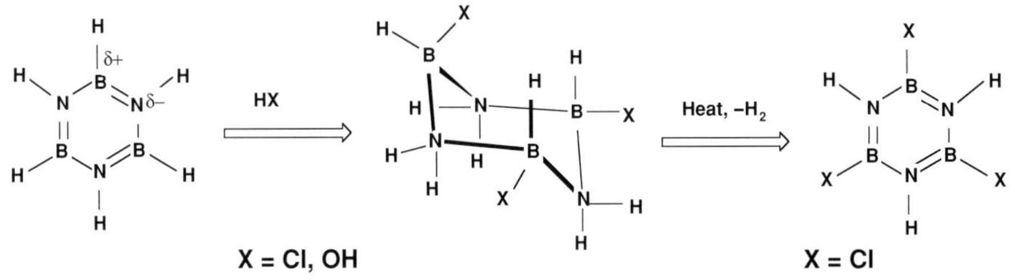

X = Cl, OH   X = Cl

Figure 4.4b Borazine is more reactive than benzene due to the polarity of the B–N bond.

Boron hydrides

The simplest hydride (with stoichiometry BH_3) is the dimeric compound **diborane**, B_2H_6. Whilst at first sight this does not seem particularly unusual, in reality it has considerable implications for the bonding theory of this and related compounds. Historically, it was assumed that it had the same structure as ethane, C_2H_6, but since it has two valence electrons fewer it was referred to as an **'electron-deficient compound'**. The structure derived from gas phase electron diffraction studies is shown in Figure 4.4c and is different from ethane. From the B−H distances it can be seen that the bridging bonding is weaker than the bonding between boron and the terminal hydrogen atoms. Any description of the bonding must account for this. In a simplified explanation of the bonding each boron atom is sp^3 hybridised. Two of these hybrid orbitals overlap with the $1s$ on each of the terminal hydrogen atoms to give conventional **two centre–two electron** (2c–2e) bonds. The remaining two sp^3 hybrid orbitals on each boron overlap with the bridging H $1s$ to give the bridging bonds. As each boron only supplies one electron to the bridge we have two **3c–2e** B−H−B bonds. This simplified bonding description is illustrated in Figure 4.4d, and accounts for the shorter 2c–2e bonds and the longer 3c–2e bonds in the bridge. Here the main principle is that maximum use is made of all the low energy orbitals – in other words each boron atom has electron density in all four of its valence orbitals. A monomeric BH_3 molecule would have a vacant *p*-orbital. A more rigorous description using LCAO would be required to explain satisfactorily the detailed electronic properties of the molecule. Thus in B_2H_6 we have exactly the correct number of electrons for the bonding mode adopted.

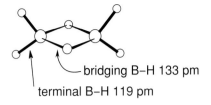

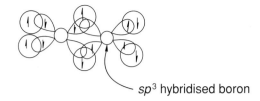

Figure 4.4c The structure of diborane.

Figure 4.4d A simplified bonding scheme for diborane.

Diborane is conveniently prepared by reaction of borohydride with iodine. It is typical of all the lower boron hydrides in that it is **thermally unstable** and **oxygen and moisture sensitive.** Cleavage of the bridge readily occurs with Lewis bases, L, to give a variety of adducts.

$$2[BH_4]^- + I_2 \rightarrow B_2H_6 + 2I^- + H_2$$
$$B_2H_6 + 3O_2 \rightarrow B_2O_3 + 3H_2O$$
$$B_2H_6 + 6H_2O \rightarrow 2B(OH)_3 + 6H_2$$
$$B_2H_6 + 2L \rightarrow L \cdot BH_3 \quad L = Me_3N, Ph_3P, H^-, \text{etc.}$$

Reaction with alkenes is of considerable importance in organic synthesis. The alkylboranes thus formed can react to give a variety of useful products, for instance with H^+ to give alkanes and with alkaline peroxide to yield alcohols, as indicated below.

$$B_2H_6 + 6RCH=CH_2 \rightarrow 2B(CH_2CH_2R)_3$$
$$B(CH_2CH_2R)_3 + H^+ \rightarrow RCH_2CH_3$$
$$B(CH_2CH_2R)_3 + OH^- + H_2O_2 \rightarrow RCH_2CH_2OH$$

A family of related larger boron hydrides and boron hydride anions is known and given the general name **boranes**; important examples are B_4H_{10}, B_5H_9, B_6H_{10}, $B_{10}H_{14}$ and $[B_{12}H_{12}]^{2-}$. These contain more than one boron atom and are termed **clusters**. At first sight their formulae and structures appear complex, but a system of classifying and rationalising the bonding within them has been developed. Structures are given general names which relate to their overall appearance: '*closo*' (closed polyhedra), '*nido*' (nest-like) and '*arachno*' (web-like). General rules (often called **Wade's rules**) relate the formula and the number of skeletal electron pairs available to the structure adopted, and are summarised below.

General formula	Number of skeletal electron pairs	Polyhedral structure	General name (descriptor)
$[B_nH_n]^{2-}$	$n+1$	n vertices of an n vertex polyhedron	*closo*
B_nH_{n+4}	$n+2$	n vertices of an $n+1$ vertex polyhedron	*nido*
B_nH_{n+6}	$n+3$	n vertices of an $n+2$ vertex polyhedron	*arachno*

In counting the number of electron pairs it is assumed that structures consist of BH units, each providing two electrons (each boron has three valence electrons and uses one to bond to H leaving two for skeletal bonding), and the remaining H atoms (whether terminal or bridging) contribute a further electron each. For example, B_4H_{10} (a member of the B_nH_{n+6} family with $n = 4$) has $4 \times$ BH contributing $8e^-$, and 6H contributing $6e^-$, making a total of $14e^-$, i.e. 7 electron pairs ($n + 3$ electron pairs). Thus we would expect the compound to adopt a structure based on an $n + 2$ vertex, i.e. a 6 vertex polyhedron (an octahedron). Useful **structural correlations** can be made for clusters with the same number of electron pairs, as shown in Figure 4.4e. These rules have been used to rationalise existing structures and predict the structures of new boranes.

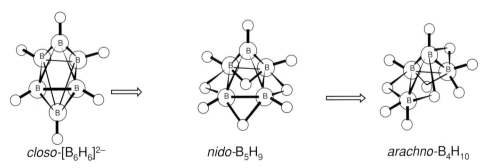

closo-$[B_6H_6]^{2-}$ *nido*-B_5H_9 *arachno*-B_4H_{10}

Figure 4.4e The relationship between *closo*, *nido* and *arachno* structures with 7 electron pairs, all based on the 6-vertex polyhedron.

4.5. Group 14 Elements (C, Si, Ge, Sn, Pb)

Overview

The existence of **allotropes** (different structural forms of the same element) of carbon has excited much recent interest with the discovery of a new form of the element. In addition to the more well established diamond and graphite structures (Figure 4.5a) the spherical molecule **buckminsterfullerine**, C_{60}, was the first of a number of related forms of carbon to be discovered in the mid 1980s.

The Group oxidation state of +4 is the most stable for C and Si but gradually becomes less so on descending the group; for Sn the +4 and +2 oxidation states are common, and for Pb the +2 state is more stable. Almost all of the compounds formed by Group 14 elements are covalent, with the exception of PbF_2 which has a typical ionic structure.

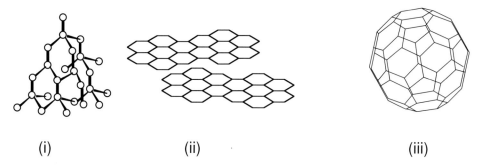

| (i) | (ii) | (iii) |

Figure 4.5a Some allotropes of carbon. (i) Diamond (Si and Ge have the same structure), (ii) graphite and (iii) buckminsterfullerine, C_{60}.

Hydrides

The hydrides are of vast importance in carbon chemistry, where their stability is in part due to the high strength of the C−C and C−H bonds. For the remaining elements the range of hydrides is very much reduced.

$$Si_nH_{2n+2} \quad n = 1 \text{ to } 8$$
$$Ge_nH_{2n+2} \quad n = 1 \text{ to } 5$$
$$Sn_nH_{2n+2} \quad n = 1, 2$$
$$PbH_4$$

One reason for this is the steady decrease in the E−E and E−H bond strengths on descending the group, with an approximately 50% drop in each from C to Pb.

Multiple bonds

As a result of the progressively **weaker π-bonding** for the **heavier elements** the unsaturated derivatives common for carbon are much less numerous for Si, Ge, Sn and Pb. Thus compounds with E=E double bonds (E = Si, Ge, Sn, Pb) can generally only be formed when sterically demanding groups are present, such as shown in Figure 4.5b.

Halides

A large range of tetrahalides are known and most have the expected tetrahedral structures. The structure EX_4 exists for all combinations of E and X, except for Pb where the +4 oxidation state is strongly oxidising and thus only PbF_4 is stable at room temperature. As is frequently found, unstable compounds can be stabilised by the formation of complexes. Thus whilst **$PbCl_4$ is unstable** at room temperature, oxidation of $PbCl_2$ with chlorine in the presence of chloride ion donors gives stable salts of $[PbCl_6]^{2-}$. With the exception of carbon, all the tetrahalides are Lewis acids and form numerous complexes with

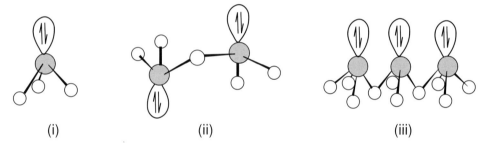

Figure 4.5b   The stabilisation of the Si=Si bond by bulky substituents. Similar compounds can be formed by Ge and Sn.

neutral or anionic ligands to give trigonal bipyramidal EX_4L, $[EX_5]^-$ and octahedral EX_4L_2, $[EX_6]^{2-}$ complexes. Again, with the exception of carbon, the tetrahalides tend to hydrolyse rapidly in the presence of water. The reason for both of these observations lies in the inability of carbon to expand its octet and accommodate a fifth group in its primary co-ordination sphere, thus CX_4L or $[CX_5]^-$ have not yet been prepared.

The dihalides are known for Si but are not stable, germanium(+2) halides can be prepared by reaction of GeX_4 with Ge, but GeX_2 disproportionates on heating.

$$2GeX_2 \rightarrow Ge + GeX_4 \qquad \text{thermal disproportionation of } GeX_2$$

The dihalides of tin and lead are stable and have a rich co-ordination chemistry giving complexes such as $[SnCl_3]^-$, $[Sn_3F_{10}]^{4-}$, $[PbCl_4]^{2-}$ and $[PbX_6]^{4-}$; the structures of some of the tin complexes are illustrated in Figure 4.5c.

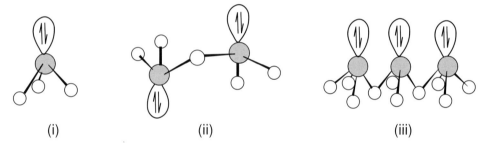

(i)   **(ii)**   **(iii)**

Figure 4.5c   The structure of some tin(II) complexes. (i) $[SnCl_3]^-$; (ii) $[Sn_2F_5]^-$; (iii) $[Sn_3F_{10}]^{4-}$. The lone pair on the tin (shaded atoms) is stereochemically active in each case.

Oxides

As a result of **strong π-bonding** between two period 2 elements, the oxides of carbon adopt molecular structures. Thus CO_2 is a monomeric linear molecule whilst other EO_2 compounds have giant lattice structures. The compound PbO_2 is a strong oxidising agent, as is to be expected of a period 6 element in its highest oxidation state. On heating it decomposes via a series of oxides such as Pb_3O_4 (best thought of as $2PbO \cdot PbO_2$), to eventually give PbO and O_2.

The existence of a stable oxide in the +2 oxidation state for carbon, carbon monoxide, is presumably a result of strong π-bonding. However, SiO is not stable, and GeO, SnO and PbO have layered structures and are stable at room temperature.

The π-bonding also accounts for differences between carbon and the other elements in the structure of their oxyanions. The carbonate ion $[CO_3]^{2-}$, formed by dissolving CO_2 in alkaline solution, has a trigonal planar arrangement about the carbon. The free acid, H_2CO_3, can be isolated at low temperature. Most metal carbonates are insoluble in water, the exception being the Group 1 salts, M_2CO_3. Carbonates decompose to metal oxide and CO_2 on heating and carbonates of small highly charged (i.e. highly polarising) ions decompose below room temperature.

$$\begin{aligned}
CaCO_3 &\rightarrow CaO + CO_2 \text{ at } 832°C \\
MgCO_3 &\rightarrow MgO + CO_2 \text{ at } 394°C \\
Al_2(CO_3)_3 &\rightarrow Al_2O_3 + 3CO_2 \text{ below } 0°C
\end{aligned}$$

Simple silicates, SiO_4^{4-} are tetrahedral at Si and a vast number of related silicate ions exist which also have tetrahedrally co-ordinated Si centres (see Section 4.6). In contrast to carbonates the silicates have high thermal stability.

4.6. Silicate and Siloxane Chemistry

Water-glass solutions

Fusion of sand (SiO_2) with Na_2CO_3 at $> 1300°C$ results in water soluble glasses, and aqueous solutions of these glasses are described as **water-glass solutions**. The composition of these solutions is complex, but equilibrium concentrations of the various silicate species present are rapidly attained at pH > 10. Mono silicate anions, e.g. $SiO(OH)_3^-$ and $SiO_2(OH)_2^{2-}$, predominate in solutions with high Na/Si ratios, and larger anionic colloidal species (1.5–5 nm diameter particle size) are present in solutions with higher silica content. The addition of a few crystals of transition metal salts to water-glass solutions results in crystallisation of metal silicates and the formation of beautiful 'crystal gardens'. At lower pH the relatively simple silicates polymerise to give more complex ions. This type of reaction **lowers the charge density** on the ions and is common to any chemical species bearing a high negative charge.

$$2SiO_4^{4-} + 2H^+ \rightarrow Si_2O_7^{6-} + H_2O$$
$$SiO_4^{4-} + 2H^+ \rightarrow SiO_3^{2-} + H_2O$$
$$SiO_4^{4-} + 4H^+ \rightarrow SiO_2 + 2H_2O$$

Silica gels are widely used as desiccants and these are readily obtained by acidification of low sodium ratio water-glass solutions.

Solid-state silicate anions

A vast number of silicate ions exist in the solid-state and these can be considered as being made by SiO_4 tetrahedra sharing corners. The simplest possible of these, the disilicate ion, $Si_2O_7^{6-}$ is made by two SiO_4^{4-} ions sharing one corner, as shown in Figure 4.6a. Cyclic anions (Figure 4.6b), single-chained and double-chained polymers, and layered structures (Figure 4.6c) are some representatives of the vast array of silicate ions found in the solid-state. These are naturally occurring in a variety of minerals. Asbestos has a double-chained silicate structure (Figure 4.6cii), whilst talc has a structure with polymeric layers, one of which is illustrated in Figure 4.6ciii. There are many subtle variations possible in the structures illustrated, for instance in Figure 4.6b the linear chain may have 'steps' present leading to a different repeating unit and hence formulae such as $Si_2O_6^{4-}$ and $Si_4O_{12}^{8-}$.

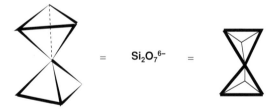

Figure 4.6a Representations of the structures of the disilicate ion.

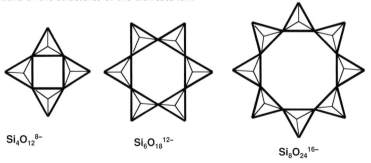

Figure 4.6b Idealised structures of some cyclic silicates.

44

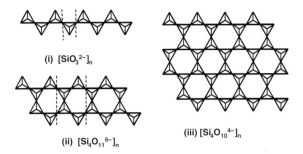

(i) $[SiO_3^{2-}]_n$

(ii) $[Si_4O_{11}^{6-}]_n$

(iii) $[Si_4O_{10}^{4-}]_n$

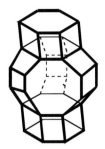

Figure 4.6c Idealised structures of (i) single- and (ii) double-chain and (iii) layered silicates. The repeating units of (i) and (ii) are indicated by the dashed lines.

Figure 4.6d The cage structure of the zeolite $Na_8Ca_4Al_8Si_{16}O_{48}$.

Zeolite chemistry

Zeolites are crystalline **aluminosilicates** with the general formula $M_a[(AlO_2)_b(SiO_2)_c]\cdot dH_2O$. They are characterised as a combination of rigid anionic networks with **channels and cavities of well-defined dimensions**, and labile charge-compensating cations. Each $\{AlO_2\}$ unit brings a negative charge to an otherwise neutral isostructural silicate structure. Approximately 40 naturally occurring zeolites are known and more than 100 synthetic zeolites have been successfully characterised. The primary building block units in zeolites are SiO_4 and AlO_4^- tetrahedra which link together by corner sharing, producing a variety of framework structures. One example, shown in Figure 4.6d, illustrates the framework structure adopted by the aluminosilicate anion. For these structures the junctions of lines represent the SiO_4 and AlO_4 tetrahedral, and counterions are located within the cage structure.

Zeolites have found applications that utilise the open spaces within the framework. Molecular sieves (which trap small molecules in their cavities), cation exchangers, and, more recently, use of their acidic properties have encouraged their large-scale industrial use as catalysts for organic transformations.

Siloxanes

Siloxanes (silicones) have the general formula $(R_2SiO)_n$ and have cyclic or polymeric structures. Their chemical inertness, exceptional thermal stability, physiological inertness and water repellency properties have been exploited in many applications, e.g. lubricants, sealants, hydrophobic coatings and medical devices.

The **oligomeric structures** of siloxanes contrast with the monomeric structures observed for the carbon containing species with the same empirical formulae (ketones) as shown in Figure 4.6e. Examination of the bond energies of the σ- and π-components of C$-$O and Si$-$O bonds is instructive and indicates that it is **thermodynamically more favourable** for Si to form two σ-bonds to two O atoms rather than the combination of σ- $+$ π-bonds to one O; see Section 4.2. The explanation for the relative bond energy values lies in a combination of factors:

(i) Si is more electropositive than C and hence the Si$-$O bond is more polar than the C$-$O bond. This leads to a greater ionic/electrostatic contribution to the σ-bond strength.

(ii) The $2p/2p$ orbital interaction for the C$-$O π-bond is significantly stronger than the analogous $3p/2p$ interaction anticipated for a Si$-$O π-bond. This is because of the mismatch of orbital size between the second and third period elements.

(iii) The Si atom has relatively low-energy $3d$-orbitals which are believed to be implicated in an additional bonding interaction to O. The interaction envisaged involves formation of a **$2p/3d$ donor–acceptor π-bond** by donation of a pair of electrons from a filled O $2p$-orbital. This type of π-bond is a fairly common stabilising influence between second and third period elements.

The presence of such $p\pi$–$d\pi$ interactions is often cited as an explanation for the linear O$-$Si$-$O angle in molecules such as $H_3SiOSiH_3$ and the planarity of $(H_3Si)_3N$. However, electrostatic repulsions arising through the polar σ-bonds are also minimised in these geometries.

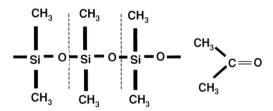

Figure 4.6e The structures of $[(CH_3)_2SiO]_n$ (dashed lines indicate the repeating unit in the polymer) and acetone $(CH_3)_2CO$.

45

4.7. Group 15 Elements – the Pnicogens (N, P, As, Sb, Bi)

Hydrides

The hydrides EH_3 are known for all the elements in Group 15 but become **less stable with increasing atomic number**. The most important is ammonia, NH_3, which is produced on a vast scale by the catalytic reaction between N_2 and H_2 (see Section 4.14). Ammonia and the related amines (R_3N), and phosphines (R_3P) have a lone pair of electrons and are important as ligands in co-ordination chemistry.

Halides

The two principal types of halide for Group 15 elements are $\mathbf{EX_3}$ and $\mathbf{EX_5}$. Nitrogen has only one stable binary halide, $\mathbf{NF_3}$, the others being **dangerously explosive**. The strength of the N–F bond stabilises NF_3, but the N–X bonds in the remaining NX_3 are sufficiently weak to make the decomposition to N_2 and X_2 thermodynamically favourable. As nitrogen is the most electronegative member of this group it is not surprising that NX_5 species are not known, and $\mathbf{[NF_4]^+}$ is the only simple nitrogen(+5) halo ion. The trihalides of the other elements are thermodynamically stable, and show some Lewis acidic properties forming complexes such as $[EX_4]^-$, where the lone pair is stereochemically active, and giving structures based on a trigonal bipyramidal distribution of electron pairs (see Section 2.2). The stability of the pentahalides alternates with phosphorus and antimony (periods 3 and 5) having some stable EX_5 compounds (X = F, Cl, Br), whilst the more oxidising arsenic(+5) and bismuth(+5) have a more limited range with only AsF_5 and BiF_5 being stable.

The tendency of the pentahalides to act both as Lewis acids and Lewis bases can be seen in the structures of some of the phosphorus halides. The compound PCl_5 exists as an ionic solid $[PCl_4]^+[PCl_6]^-$, and in non-polar solvents and the gas phase it exists as a trigonal bipyramidal monomeric species. Salts containing the cation or anion alone can be made by treating solutions of PCl_5 with a chloride acceptor or donor, respectively. PBr_5 exists as $[PBr_4]^+[Br]^-$ in the solid state, and $[PBr_6]^-$ or $[PI_6]^-$ species are not known. The halides are readily hydrolysed by water to the corresponding acids.

$$PCl_3 + 3H_2O \rightarrow H_3PO_3 \text{ (phosphorus acid)} + 3HCl$$
$$PCl_5 + 4H_2O \rightarrow H_3PO_4 \text{ (phosphoric acid)} + 5HCl$$

Oxides

Whilst P, As, Sb and Bi have the expected oxides $\mathbf{E_2O_3}$ and $\mathbf{E_2O_5}$, nitrogen has oxides in formal oxidation states from 1 to 5, as shown in Figure 4.7a. These are stable as a result of the **strong π-bonding** between N and O. Both NO and NO_2 have unpaired electrons and are paramagnetic. Nitrogen dioxide, NO_2 is in equilibrium with its dimer, N_2O_4. Nitrous oxide, N_2O supports combustion as it decomposes to N_2 and O_2 at elevated temperatures, and reacts with $NaNH_2$ to produce NaN_3. The monoxide, NO, reacts with air to give NO_2 and forms numerous complexes with transition metals which have some analogies to those of CO. Its main use is in the manufacture of nitric acid (Section 4.14). Oxides of phosphorus are shown in Figure 4.7b.

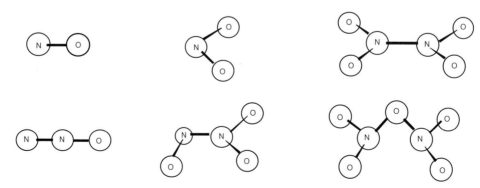

Figure 4.7a The structures of the principal nitrogen oxides. There are long N–N bonds in N_2O_3 and N_2O_4.

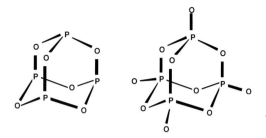

Figure 4.7b The relationship between the structures of phosphorus(+3) and phosphorus(+5) oxides.

They can be considered as being derived from the structure of P_4, with both P_4O_6 and P_4O_{10} having a residual tetrahedral arrangement of phosphorus atoms, with the phosphorus atoms themselves having a tetrahedral distribution of electron pairs. The oxides of arsenic, antimony and bismuth display an increasing tendency towards **metallic type structures**, and discrete molecular forms are not observed. The trend from non-metals to metals is evident in the chemical properties of the oxides as well as their structures. Thus, whilst all E_2O_5 are acidic, E_4O_6 are acidic (E = P), **amphoteric** (E = As, Sb) and basic (E = Bi).

Oxyanions

Of the oxyanions the trigonal planar nitrate, $[NO_3]^-$ and tetrahedral phosphate, $[PO_4]^{3-}$ are the most important. The differing formulae and structures of the ions result from the strong π-bonding between N and O. There are a large number of polyphosphoric acids and polyphosphate ions (Figure 4.7c) which are formally derived from phosphate. They all have phosphorus in a tetrahedral environment, and are formed by thermal dehydration reactions of the monomeric species.

$$H_2PO_4^- \rightarrow [H_2P_2O_7]^{2-} \rightarrow [P_3O_9]^{3-} \rightarrow [PO_3^-]_n$$

Polyphosphates form the basis of energy storage and liberation in many living systems. The hydrolysis of adenosine triphosphate (ATP) releases energy to the body, whilst energy is stored by its production.

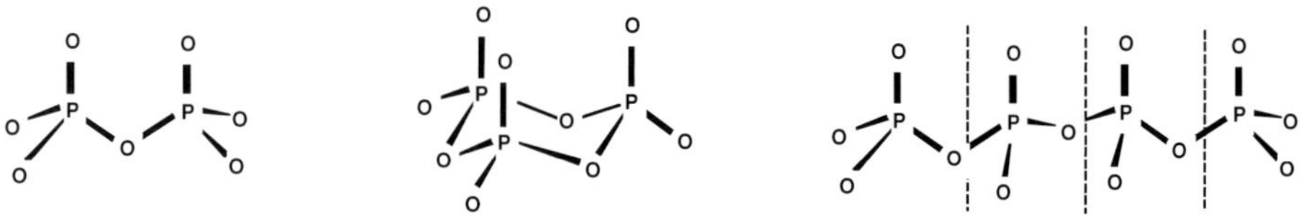

(i)   (ii)   (iii)

Figure 4.7c The structures of some polyphosphate ions. (i) $[P_2O_7]^{4-}$; (ii) $[P_3O_9]^{3-}$; (iii) $[PO_3^-]_n$ (dashed lines represent the repeating unit of the polymer).

Polyphosphazenes

Phosphorus forms a series of compounds similar to the borazines discussed in Section 4.4. Reaction of ammonium chloride with phosphorus pentachloride or bromide gives **polyphosphazenes** $(NPX_2)_n$. The product mixture contains cyclic ($n = 3, 4$) and polymeric species. The six-membered ring is planar and analogies have been made between its structure (Figure 4.7d) and that of benzene. The PN bond distances are all equal and the double bond character possibly arises from $p\pi$–$d\pi$ overlap between filled p-orbitals on N and empty $3d$-orbitals on P, as indicated in Figure 4.7e. However, the P atom must use two orthogonal d-orbitals to π-bond with its neighbouring N atoms, and because of this the π-system is not fully delocalised around the ring. The π-clouds are centred as 'islands' on each N with nodes at each phosphorus. The halides on phosphorus are readily substituted through nucleophilic attack by reaction with suitable metal salts, or by solvolysis of the P−X bond.

$$(NPX_2)_3 + MY \rightarrow (NPY_2)_3 + MX \qquad (Y = NCS, N_3, \text{alkyl, aryl, alkoxy, etc.})$$

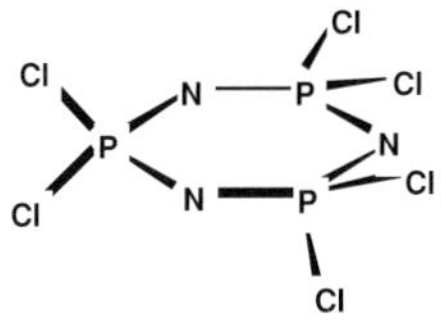

Figure 4.7d The structure of $(NPCl_2)_3$.

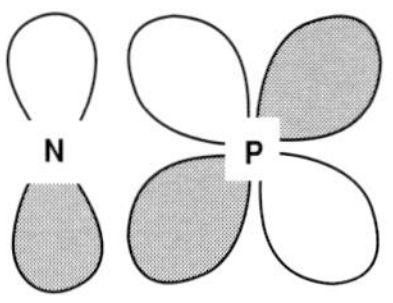

Figure 4.7e The mechanism of π-bonding in phosphazenes.

4.8. Group 16 Elements – the Chalcogens (O, S, Se, Te, Po)

Oxidation states

As oxygen is one of the most electronegative elements it is not surprising the range of oxidation states in its compounds is limited. In the vast majority of its compounds it is formally present as -2, with -1 in H_2O_2 and the peroxide ion, $[O_2]^{2-}$, and -0.5 in the superoxide $[O_2]^-$ ion also known. Oxidation states higher than zero can only be produced by the strongest of oxidants such as PtF_6 to give $[O_2]^+$, and fluorine to give O_2F_2 and OF_2 containing oxygen in oxidation states 0.5, 1 and 2 respectively. Oxygen fluorides are gaseous, unstable molecules. The difluoride, OF_2, is explosive and O_2F_2 decomposes above $-50°C$.

The -2 oxidation state for the group is represented by E^{2-} ions. The stability of these to oxidation decreases from O to Po. They react with water to form EH^- and liberate EH_2 with acids.

Oxidation states up to $+6$ are well established for S, Se and Te. Stable oxides EO_2 and EO_3 are known. Sulphur dioxide, SO_2, is a gas with the expected angular structure; SeO_2 and TeO_2 are polymeric solids, with the structure of SeO_2 shown in Figure 4.8a. Selenium dioxide dissolves in water to give the strongly acidic and oxidising H_2SeO_3. The structures of the trioxides show a transition to more metallic behaviour upon descending the Group. Whilst SO_3 exists as either a monomer, trimer or linear polymer, and SeO_3 has a trimeric structure, TeO_3 has a three-dimensional lattice structure that is typical of metals. The trimeric structures of SO_3 and SeO_3 are shown in Figure 4.8b. As expected from period 4 and 6 elements **Se(VI) and Po(VI) are strongly oxidising**; they have high reduction potentials: SeO_4^{2-}/H_2SeO_3 $E^0 = 1.15$ V and PoO_3/PoO_2 $E^0 = 1.52$ V, with SeO_3 decomposing to SeO_2 and oxygen.

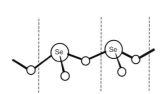

Figure 4.8a The solid-state structure of SeO_2 (dashed lines indicate the repeating unit).

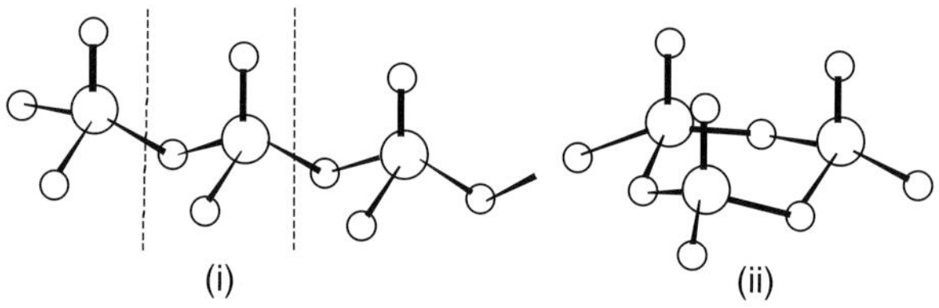

(i) (ii)

Figure 4.8b Two of the structures adopted by SO_3: (i) the linear polymer and (ii), the trimer. SeO_3 has a similar trimeric structure.

Hydrides

The hydrides EH_2 are known for all the elements, although the stability decreases markedly with atomic weight of E; H_2Te decomposes above $0°C$. Water is unique in being a liquid under ambient conditions due to the **strong hydrogen bonding** as indicated in Figure 4.8c (see Section 5.2); this being stronger than the corresponding interactions in other EH_2 as a result of

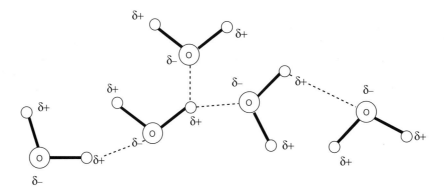

Figure 4.8c Hydrogen bonding (dashed lines) in water is strengthened by the strongly polar O—H bond.

the high electronegativity of O and hence the strongly polar O−H bond. One notable feature of all EH_2 (with the exception of water) is their vile smell, which becomes more offensive with increasing molecular mass.

		Boiling point temperatures of EH_2 (1 atm)			
E	O	S	Se	Te	Po
Boiling temp./K	373	212	232	271	310

Halides

Numerous halides exist, the most important being EX_2, EX_4 and EX_6, although not all possible combinations of E and X are known. The difluorides are unstable and have only been observed for O, S and Se. Other EX_2 are known with ECl_2 and EBr_2 for S, Se, Te, Po, and EI_2 for Po, although not all can be isolated as solid compounds. Similarly, EX_4 are widely represented with EF_4 for all E, ECl_4 and EBr_4 for Se, Te, Po, and EI_4 for Te and Po. There are interesting variations in structure, some being in accord with VSEPR theory, whilst others are not. Examples are shown in Figure 4.8d where the lone pair is **stereochemically active** [(i) and (ii)] and where it is not [(iii)]; EX_6 are only known for X = F. Their reactivity varies: SF_6 is inert to water whilst both SeF_6 and TeF_6 are hydrolysed. The difference in reactivity illustrates the importance of the availability of a **low-energy pathway** for a reaction. In SF_6 there is no room around the sulfur for water to bind as the first stage in the hydrolysis. See Section 4.5 for a similar situation in the stability of carbon halides. The hydrolysis reaction is a common feature in the halides of non-metals and proceeds to give the corresponding oxide and HX; although with limited amounts of water intermediate species can be isolated.

$$SF_4 + H_2O \rightarrow SOF_2 + 2HF$$
$$SF_6 + 3H_2O \rightarrow SO_3 + 6HF$$

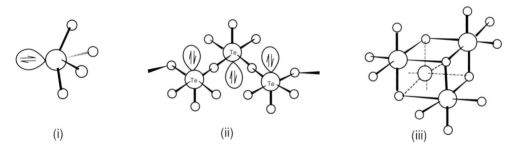

(i) (ii) (iii)

Figure 4.8d The structures of some EX_4. (i) SF_4 and SeF_4 have the 'see-saw' structure, (ii) TeF_4 is a polymeric chain and (iii) $SeCl_4$, $SeBr_4$ and $TeBr_4$ are tetramers. Note that in (i) and (ii) the lone pairs are stereochemically active, whilst in (iii) it is not, but is presumably located in a symmetrical s-orbital.

Lewis acid/Lewis base behaviour

The tetrahalides can behave as Lewis acids or bases, forming cations $[EX_3]^+$ by halide abstraction or anions on accepting halide to give $[EX_5]^-$ and $[EX_6]^{2-}$ for E = Te and Po.

An important feature of many of the compounds in Group 16 is the presence of **lone pairs** of electrons which enable the molecules to act as **donors (ligands) to metals**. This is particularly important for EH_2 and ER_2 which have rich co-ordination chemistries. Without doubt water is the most important ligand as it is the co-ordination of water to metal ions and their counterions which is responsible for the solubility of many compounds in water. Another important class of oxygen donor is the crown ethers, which have cavities capable of selectively co-ordinating to metals depending on their ionic radii (see Section 3.1).

4.9. Sulfur and its Compounds (I)

Minerals and production

Sulfur is found in nature in the form of the element itself, in reduced forms as sulfides such as FeS_2 and H_2S, and as oxidised forms in sulphate minerals such as gypsum, $CaSO_4$. Sulfur used to be mined as the element, but nowadays the main production is as a by-product of the petrochemicals industry where it is obtained by controlled oxidation of the H_2S present in natural gases. The reaction proceeds in two steps; the oxidation to sulfur dioxide followed by reaction of SO_2 with excess hydrogen sulfide to give sulfur.

$$2H_2S + 3O_2 \rightarrow 2SO_2 + 2H_2O$$
$$2H_2S + SO_2 \rightarrow 3S + 2H_2O$$
$$\text{Overall: } 2H_2S + O_2 \rightarrow 2S + 2H_2O$$

Allotropes

Sulfur forms a large number of allotropes. This is mainly due to its high propensity for **catenation**, the formation of chains, which is a feature in other aspects of its chemistry. Some of the allotropes, $S_2(g)$ and $S_3(g)$, are analogous to the oxygen allotropes but are stable only at high temperatures. Cyclic forms such as S_n ($n = 6$ up to 20) are known, often existing in different structural forms. The most stable is S_8 whose structure is shown in Figure 4.2b. Some properties are shown in Figure 4.9a. Elemental sulfur is highly reactive, combining with most elements at high temperature.

S_6	orange-red	decomposes above 0°C
S_7	intense yellow	polymerises at 45°C
S_8	pale yellow	most stable form of sulfur
S_9	intense yellow	decomposes above room temperature
S_{10}	pale yellow-green	decomposes above 0°C
S_{20}	pale yellow	decomposes on melting at 124°C

Figure 4.9a The properties of some cyclic allotropes of sulfur.

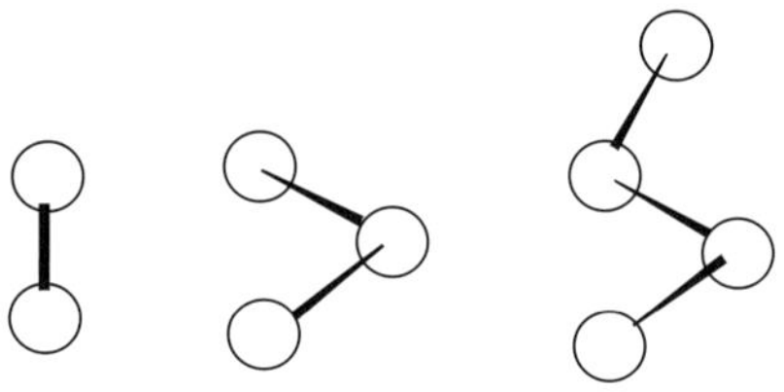

Figure 4.9b The structures of the polysulfide anions, S_2^{2-}, S_3^{2-} and S_4^{2-}.

Hydrides and sulfide anions

Because of the tendency to **catenation** there are extensive series of **sulfur hydrides** and **polysulfide anions**, H_2S_n and $[S_n]^{2-}$. The hydrides are known for $n = 2$–8 and there are a variety of preparations generally involving low temperature reactions, e.g.

$$S_nCl_2 + 2H_2S_m \rightarrow H_2S_{(n+2m)} + 2HCl \text{ (for } n + 2m = 6 \text{ to } 18)$$

All are readily oxidised by air and are unstable with respect to **disproportionation** to sulfur and H_2S.

$$H_2S_n \rightarrow H_2S + (n-1)S$$

The most stable hydride is H_2S which is moderately soluble in water (up to 0.1M) and is weakly acidic ($pK_a = 6.88$). Polysulfide anions formally derived from the hydrides are known for $n = 2$–6 and are more stable than the hydrides. The structures of some are shown in Figure 4.9b.

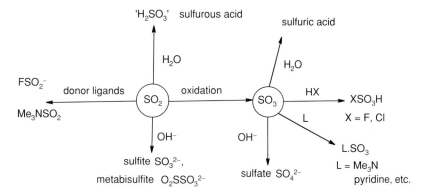

Figure 4.9c Some reactions of SO_2 and SO_3.

Oxides

Several oxides of sulfur are known. The most important of which are **SO_2** and **SO_3**; others such as SO, S_2O and S_2O_2 are unstable. When sulfur is burnt in air, SO_2 is the main product with small amounts of SO_3. Some reactions of SO_2 and SO_3 are shown in Figure 4.9c. Sulpur dioxide is manufactured industrially by the roasting of sulfide ores in air, e.g.

$$4FeS_2 + 11O_2 \rightarrow 2Fe_2O_3 + 8SO_2$$

Sulfur dioxide is also formed by numerous industrial processes and has caused considerable environmental concern as it is one of the primary causes of acid rain. It is a colourless gas which has a choking odour. It is mildly reducing and is oxidised by high oxidation state transition metals such as MnO_4^- and VO_3^-. It is considerably **more reducing in alkaline solution** as the reduction potentials below indicate:

$$SO_4^{2-} + 4H^+ + 2e^- \rightarrow SO_2 + 2H_2O \qquad E^0 = 0.16\,V \text{ at pH } 0$$
$$SO_4^{2-} + H_2O + 2e^- \rightarrow SO_3^{2-} + 2OH^- \qquad E^0 = -0.94\,V \text{ at pH } 14$$

Whilst it is very soluble in water, giving weakly acidic solutions, the expected sulfurous acid, H_2SO_3, has never been isolated. In aqueous solution the main reaction appears to be:

$$SO_2 + 2H_2O \rightarrow H_3O^+ + HSO_3^-$$

Interestingly, crystallographic evidence suggests that the proton in the hydrogen sulfite ion is bonded to the S rather than the more electronegative oxygen atom, whilst solution measurements indicate that there is an equilibrium between O- and S-bonded forms. Salts of H_2SO_3, the sulfites, are well known and stable, and are formed on dissolving SO_2 in alkaline solution. In the presence of excess SO_2 the metabisulfite ion, $S_2O_5^{2-}$ is formed.

Both SO_2 and the sulfite ion have lone pairs of electrons on S and O atoms and can form complexes with metals, bonding through either S or O depending on the electronic properties of the metal. **Hard metals**, i.e. those with high oxidation states and low polarisability, tend to bind electrostatically with **hard O-donors** and will be co-ordinated by the O of SO_2. On the other hand more polarisable metal ions in low oxidation states (**soft metals**) tend to form complexes with **softer S-donor** atoms and will be co-ordinated by the S of SO_2, as illustrated in Figure 4.9d.

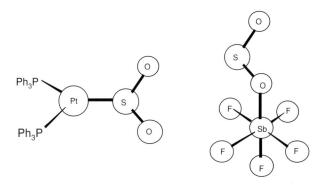

Figure 4.9d Co-ordination of SO_2 to 'soft' Pt(0) and 'hard' Sb(V) metal centres.

4.10. Sulfur and its Compounds (II)

Oxyanions and related compounds

The most commercially important reaction of SO_2 is its oxidation to form sulfuric acid (**contact process**). The reaction between SO_2 and O_2 is slow and a vanadium pentoxide catalyst at $700°C$ is used. The sequence of reactions is oxidation of SO_2 to SO_3 by V_2O_5 followed by regeneration of the catalyst by air oxidation:

$$SO_2 + V_2O_5 \rightarrow 2VO_2 + SO_3 \quad \text{and} \quad 2VO_2 + \tfrac{1}{2}O_2 \rightarrow V_2O_5$$

The SO_3 produced is dissolved in concentrated sulfuric acid rather than water, as the latter has a higher vapour pressure and causes the formation of a sulfuric acid mist in the reactor which leads to technical problems. Sulfur trioxide is a strong oxidant and a Lewis acid, forming stable complexes with electron pair donors such as organic bases and fluoride and chloride ions, as shown in Figure 4.9c.

The sulfate ion has the expected tetrahedral structure, and condensed sulfates such as disulfate, $S_2O_7^{2-}$ and trisulfate, $S_3O_{10}^{2-}$ are known forming an interesting structural parallel with chromium(VI) chemistry in the chromate, dichromate, trichromate ions.

Sulfur forms many other oxyacids and anions. One of the most important is the **thiosulfate** ion, $S_2O_3^{2-}$ which can be thought of as being formally derived by replacing one of the oxygen atoms in the sulfate ion with a sulfur. It is prepared by boiling an aqueous solution of a sulfite with sulfur, and whilst the free acid is not stable in water it can be prepared in its absence by reaction of chlorosulfonic acid with hydrogen sulphide:

$$SO_3^{2-} + S \rightarrow S_2O_3^{2-} \quad \text{and} \quad ClSO_3H + H_2S \rightarrow H_2S_2O_3 + HCl$$

The largest **commercial use of thiosulfate** is in photography, where it is used to dissolve silver bromide, forming the $[Ag(S_2O_3)_2]^{3-}$ complex ion in which the thiosulfate bonds to the soft Ag^+ centre through the terminal sulfur atom. Thiosulfate is used in the analytical determination of iodine where it is quantitatively and rapidly converted to the tetrathionate ion, $S_4O_6^{2-}$.

$$2S_2O_3^{2-} + I_2 \rightarrow S_4O_6^{2-} + 2I^-$$

The structures of the thiosulfate and tetrathionate ions are shown in Figure 4.10a.

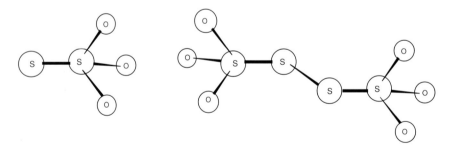

Figure 4.10a The structures of the thiosulfate, $S_2O_3^{2-}$ and tetrathionate, $S_4O_6^{2-}$ ions.

There exist a host of ions and free acids related to the tetrathionate ion in which SO_3^- groups are linked by increasing numbers of sulfur atoms, again emphasising the tendency of sulfur to **catenate**. For example, in the series of free acids $HO_3SS_nSO_3H$, compounds are known for n up to 12. The free acids are generally not stable in aqueous solution but can be prepared in anhydrous media. Crystalline salts, the so-called 'polythionates', $[O_3SS_nSO_3]^{2-}$ are more stable. Their formation and chemistry is complicated by competing redox and catenation processes. There are a number of synthetic routes to polythionates and the dithionate ion, $n = 0$, can be prepared by selective oxidation of SO_2 by MnO_2.

$$SCl_2 + 2HSO_3^- \rightarrow [O_3SSSO_3]^{2-} + 2HCl$$
$$2MnO_2 + 3SO_2 \rightarrow MnSO_4 + MnS_2O_6$$

Redox chemistry

The **redox chemistry** in aqueous solution is conveniently summarised in the Latimer diagram (see Section 1.5) shown in Figure 4.10b. From this, the instability of many of the oxidation states to **disproportionation** can be seen. Thus it can immediately be deduced that $S_2O_6^{2-}$, $S_3O_6^{2-}$ and $S_2O_3^{2-}$ are unstable to disproportionation to adjacent oxidation states. Indeed a closer inspection reveals that only HSO_4^-, S and H_2S are stable in acidic solution. The fact that many of these sulfur species are observable in aqueous solution is due to their **kinetic stability**.

Sulfur also has a significant chemistry associated with the peroxo group (O_2^{2-}) and forms acids and related salts such as K_2SO_5 (peroxomonosulfate) and $K_2S_2O_8$ (peroxodisulfate), the structures of which are shown in Figure 4.10c. At first sight these contain sulfur in the +8 and +7 oxidation states respectively. However, these are impossibly high for a Group 16 element and the explanation lies in the unusual oxidation state of some of the oxygen atoms present. In the peroxo group, oxygen has a formal oxidation state of -1. The free acids of both have been obtained, but peroxomonosulfuric acid is dangerously explosive, and its salts are unstable. Peroxodisulfuric acid and its salts are more stable and are very strong oxidising agents.

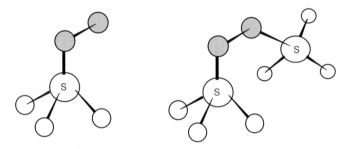

$$HSO_4^- \xrightarrow{-0.25\ V} S_2O_6^{2-} \xrightarrow{0.57\ V} H_2SO_3 \xrightarrow{0.26\ V} S_3O_6^{2-} \xrightarrow{0.73\ V} S_4O_6^{2-} \xrightarrow{0.16\ V} S_2O_3^{2-} \xrightarrow{0.47\ V} S \xrightarrow{0.14\ V} H_2S$$

0.16 V   0.45 V

Figure 4.10b   A Latimer diagram for the oxidation states of sulfur in acidic solution.

Figure 4.10c   The structures of the peroxomonosulfate, SO_5^{2-} and the peroxodisulfate, $S_2O_8^{2-}$ ions. The peroxo oxygen atoms are shaded.

Sulfur–nitrogen compounds

Sulfur has extensive chemistry with nitrogen forming binary compounds of which S_4N_4 is the most important. It can be prepared by reaction of S_2Cl_2 with NH_4Cl:

$$6S_2Cl_2 + 4NH_4Cl \rightarrow S_4N_4 + S_8 + 16HCl$$

It is an endothermic compound and can be treacherously explosive, detonating unpredictably on rapid heating or when struck.

Compound S_4N_4 is stable to water but hydrolysed by base to $S_2O_3^{2-}$, NH_3 and, depending on the conditions, either $S_3O_6^{2-}$ or SO_3^{2-}. When the vapour is passed over heated silver wool at low pressure, ring contraction occurs and N_2S_2 can be obtained. This is even less stable than S_4N_4, being shock sensitive and explosive above room temperature. It slowly polymerises to give $(SN)_n$ which is reasonably stable but will explode at elevated temperatures. There has been considerable interest in this latter compound as it is a superconductor at low temperature. The structures of S_4N_4, S_2N_2 and $(SN)_n$, the latter two being essentially planar, are shown in Figure 4.10d. The bonding in S_4N_4 is not fully understood and there are two long S.....S bonds across the cage.

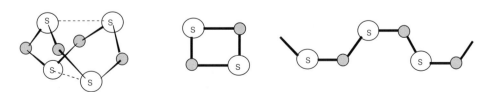

Figure 4.10d   The structures of S_4N_4, S_2N_2 and $(SN)_x$.

53

4.11. Group 17 Elements – the Halogens (F, Cl, Br, I, At)

Overview

The halogens exist as diatomic molecules, $F_2(g)$, $Cl_2(g)$, $Br_2(l)$ and $I_2(s)$. The isolated atoms have an ns^2np^5 electronic configuration and hence the most stable oxidation state for all of them is -1, as found in the halide anions. The range of oxidation states displayed by the halogens increases on descending the group. Fluorine is the **most electronegative element** in the Periodic Table and its chemistry is restricted to the zero and -1 oxidation states in F_2 and F^-, and formally $F(-1)$ in all of its other compounds. The 'oxide' F_2O is better thought of as an oxygen difluoride. The element is difficult to prepare chemically (Figure 4.11a), and is normally produced by **electrolysis** of molten KHF_2/HF. The exceptionally high reactivity of F_2 can be attributed to a weak $F-F$ single bond and to the strong bonds formed between F and other elements. The data in Figure 4.11b emphasise that the bonding in F_2 is much weaker than might be expected. The particularly weak $F-F$ bond is thought to arise as a consequence of the inter-nuclear repulsions and lone-pair/lone-pair repulsions that would occur at the expected bond distance of 128 pm (Figure 4.11c). All the other halogens are readily prepared by oxidation of their halide ions.

$$K_2MnF_6 + 2SbF_5 \longrightarrow 2KSbF_6 + MnF_3 + \tfrac{1}{2}F_2$$

Figure 4.11a The chemical synthesis of fluorine. Whilst the $[MnF_6]^{2-}$ ion is stable, MnF_4 is not. The reaction can be considered as fluoride abstraction from $[MnF_6]^{2-}$ followed by decomposition of MnF_4.

X	$2\times$ covalent radius / pm	observed X–X distance / pm	Bond dissociation energy/ kJmol^{-1}
F	128	143	158
Cl	198	199	242
Br	228	228	193
I	266	266	151

Figure 4.11b Some physical properties of the molecular halogens, X_2.

Oxidation states

The -1 oxidation state is the most stable for all the halogens. The ease of oxidation increases along the series $F^- < Cl^- < Br^- < I^-$ and thus the heavier halide ions react with lighter halogens for example:

$$2I^- + Br_2 \rightarrow I_2 + 2Br^-$$

Higher oxidation states of chlorine can only be produced in combination with more electronegative elements, namely fluorine and oxygen. Compounds in oxidation states up to the theoretical maximum of $+7$ are known, the commonest examples being represented by the oxides and the more stable oxyanions derived from them.

Oxidation state	$+1$	$+3$	$+4$	$+5$	$+6$	$+7$
Oxide	Cl_2O	Cl_2O_3	ClO_2	Cl_2O_5	Cl_2O_6	Cl_2O_7
Oxyanion	ClO^-	ClO_2^-		ClO_3^-		ClO_4^-

All the oxides are **thermodynamically unstable** with respect to oxygen and chlorine, and the oxyanions are all unstable with respect to **disproportionation** to Cl^- and ClO_4^-.

The bond distances in the ClO_x^- ions decrease with increasing x reflecting the increasing resonance stabilisation of the anions shown in Figure 4.11d, which, along with possible steric effects, also accounts for their increasing **kinetic stability**. The range of oxides and oxyanions of bromine and iodine is less extensive. **Disproportionation** of the intermediate oxidation states is common for Cl, Br and I, with the **rate of reaction increasing** with atomic number. Whilst ClO^-, BrO^- and ClO_2^-

can be readily obtained, IO^-, BrO_2^- and IO_2^- have never been observed. As expected (see Section 4.1) bromine($+7$) can only be obtained with difficulty: BrO_4^- being produced by oxidation of BrO_3^- with strong oxidants such as XeF_2 or F_2. This is thermodynamically the strongest oxidising agent of the EO_4^- anions, but in common with ClO_4^- its reactions are often very slow due to the strong halogen–oxygen bonds. Whilst tetrahedral geometries are found for all EO_4^- ions, the larger iodine is capable of accommodating six oxygens to give the octahedral IO_6^{5-} and the related H_5IO_6.

The trend in the thermodynamic stability of the EO_3^- anions reflects the instability of the high oxidation states of period 4 elements, with BrO_3^- being the least and IO_3^- the most stable. Thus, iodine will displace bromine or chlorine from bromates and chlorates respectively.

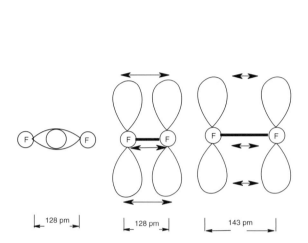

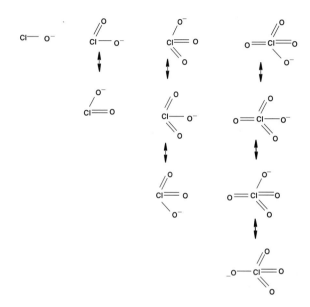

Figure 4.11c Bonding in F_2. At 128 pm the σ-bonding is optimised, but the lone-pair/lone-pair repulsions and the repulsions between the positively charged nuclei are high. These unfavourable interactions are reduced at the observed bond distance of 143 pm, which represents a balance between the various interactions.

Figure 4.11d Resonance structures of the $[ClO_x]^-$ ions. Bond distances decrease as the average Cl–O bond order increases from 1 in $[ClO]^-$ to 1.75 in $[ClO_4]^-$.

Interhalogen compounds and polyhalides

There are a considerable number of **interhalogen compounds**, XY_n, where X is the heavier halogen. The range of compounds decreases as the oxidation states increase. Thus, all EX are known, but only EF_3 and ICl_3 are known for E($+3$). For the higher oxidation states only EF_5 and only IF_7 exist. These compounds have Lewis acid/base properties and can donate or accept F^- giving $[EX_{n-1}]^+$ and $[EX_{n+1}]^-$ ions respectively, with their structures conforming to VSEPR rules (Section 2.2).

Molecular halogens can accept halide ions to give so-called **polyhalides**. The most important of these are the polyiodides, I_n^-, where $n = 3, 5, 7$, etc. The **triiodide ion** forms an intense blue coloured complex when starch is added to I_2/I^- solutions in volumetric analysis. A wide range of salts $[XYZ]^-$ can be isolated which have the heaviest halogen as the central atom. Thermal decomposition of these salts leaves the metal halide with the highest lattice energy. Thus K[ClIBr] on heating decomposes to KCl and IBr.

Donor properties of halide anions

Halide ions have four lone pairs with which to bind as ligands to metals. A vast number of halide complexes are known. Larger metals in lower oxidation states (so-called '**soft metals**') such as Pt($+2$) and Hg($+2$) form strong complexes with iodide (the 'softest' of the halides) such as $[HgI_4]^{2-}$ and $[PtI_4]^{2-}$. Fluoride tends to form its strongest complexes with high oxidation state species (the '**hard**' **acceptors**) where the bonding has an appreciable ionic component, such as $[TiF_6]^{2-}$. The combination of I^- with high oxidation state metal ions is rare as it is readily oxidised to I_2, although in some instances the formation of the complex ion modifies the redox properties of the metal rendering the complex unexpectedly stable, for example in the $[PtI_6]^{2-}$ ion (see Sections 1.5 and 4.3).

4.12. Group 18 Elements – the 'Inert' Gases (He, Ne, Ar, Kr, Xe, Rn)

Introduction

This family of elements possesses a closed outer electron shell, $1s^2$ for helium and ns^2np^6 for the remainder. As a result, for many years it was thought that there would be no chemistry of these elements. In 1962 PtF_6 was found to oxidise molecular oxygen to give the O_2^+ cation. It was noticed that the ionisation energy of O_2 was higher than that of Xe and thus PtF_6 should also oxidise xenon. It did and this discovery opened the field of Group 18 chemistry. Since then compounds have been identified for all except the lightest elements, He and Ne. Compounds of Xe in oxidation states from 0 to +8 have been characterised, some examples are shown in Figure 4.12a. In many cases the structures of the compounds formed were successfully predicted by VSEPR theory before they were determined experimentally.

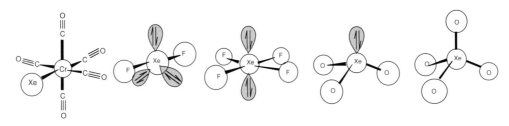

Figure 4.12a Xenon compounds showing the element in oxidation states 0, 2, 4, 6 and 8.

Xenon fluorides

Xenon reacts with elemental fluorine to form three binary fluorides, depending on reaction conditions.

$$Xe(g) + F_2(g) \rightarrow XeF_2(s)\,,\, XeF_4(s), XeF_6(s)$$

Of these **XeF_2** is the most stable and does not undergo decomposition in aqueous solution at 0°C. It is very strongly oxidising as illustrated by the reduction potential:

$$XeF_2 + 2H^+ + 2e^- \rightarrow Xe + 2HF \qquad E^0 = 2.64\,V$$

Hence, XeF_2 should oxidise water (Figure 1.5a) and thus the observed stability reflects the slowness of reaction – **kinetic stability**. Both XeF_4 and XeF_6 are much more reactive than XeF_2, and react with glass to form the dangerously explosive XeO_3. As a result they have to be handled in metal apparatus. They are both violently hydrolysed by water. These reactions are summarised in Figure 4.12b.

All the fluorides act as Lewis bases with fluoride ion acceptors, e.g. SbF_5, by donating F^- and yielding cations such as XeF^+, $Xe_2F_3^+$ (Figure 4.12c), XeF_3^+ and XeF_5^+. Only XeF_4 and XeF_6 have acceptor properties giving XeF_5^- and XeF_7^- and XeF_8^{2-} anions.

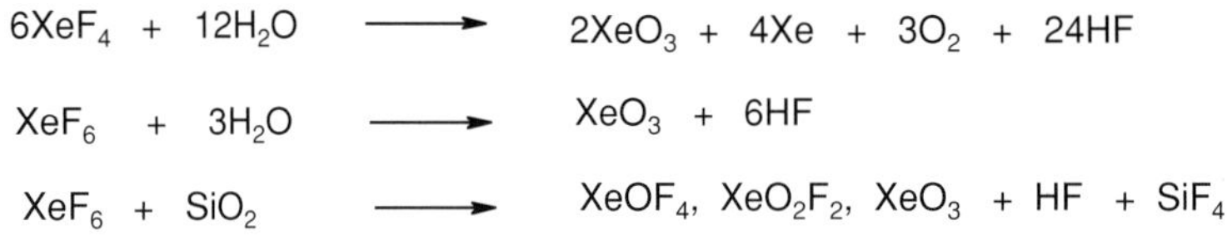

$$6XeF_4 + 12H_2O \longrightarrow 2XeO_3 + 4Xe + 3O_2 + 24HF$$

$$XeF_6 + 3H_2O \longrightarrow XeO_3 + 6HF$$

$$XeF_6 + SiO_2 \longrightarrow XeOF_4, XeO_2F_2, XeO_3 + HF + SiF_4$$

Figure 4.12b The hydrolysis of XeF_4 is complex, proceeding via combined hydrolysis of Xe–F bonds and disproportionation of Xe(+4) to Xe(0) and Xe(+6). XeF_6 reacts with glass to produce mixed oxide–fluorides and eventually XeO_3.

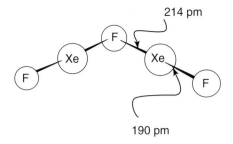

Figure 4.12c The structure of the $[Xe_2F_3]^+$ ion showing the longer bonds to the bridging fluoride.

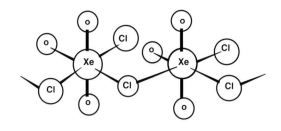

Figure 4.12d The structure of the polymeric $[XeO_3Cl_2]^{2-}$ ion; the only compound with an Xe–Cl bond that is stable at room temperature.

Xenon oxygen compounds

Compound XeO_3 acts as a Lewis acid towards fluoride and chloride ions, where XeO_3F^- and $XeO_3Cl_2^{2-}$, respectively, can be isolated. The latter being the only example of a Xe–Cl compound which is stable at room temperature. Its structure is shown in Figure 4.12d. Reaction of XeO_3 with hydroxide gives the **xenate ion**, $XeO_3(OH)^-$ from which stable salts can be isolated. Solutions of xenates slowly disproportionate/oxidise water to give the **perxenate ion**:

$$2XeO_3(OH)^- + 2OH^- \rightarrow XeO_6^{4-} + Xe + O_2 + 2H_2O$$

Salts can be isolated from these solutions and the structure of the perxenate ion is found to be a regular octahedron, as expected. Careful acidification of these salts gives XeO_4, a dangerously explosive gas.

$$Ba_2XeO_6 + 2H_2SO_4 \rightarrow XeO_4 + 2BaSO_4 + 2H_2O$$

Compounds with bonds between xenon and other elements

Xenon can also form bonds to other elements although the resulting compounds are less stable. The dichloride has been observed at low temperature but decomposes on warming. Bonds to nitrogen can be formed by reaction between XeF_2 and $HN(SO_2F)_2$ to give $FXeN(SO_2F)_2$ and $Xe[N(SO_2F)_2]_2$. Here reaction is favourable as the other product (HF) is particularly stable and the $N(SO_2F)_2$ moiety of the product is resistant to oxidation by Xe(+2). Similar approaches have led to the isolation of $Xe(OTeF_5)_4$ and $Xe(OTeF_5)_6$ by reaction of XeF_4 and XeF_6 with the oxidation-resistant $B(OTeF_6)_3$. As xenon has four lone pairs of electrons it might be expected that the element itself should be capable of acting as a ligand to metals, and such behaviour has indeed been found. Elemental Xe can co-ordinate to $Cr(CO)_5$ generated photolytically from $[Cr(CO)_6]$. However, the resulting complex is only observable at low temperature and has a half-life of about 2 s at $-98°C$. Complexes such as $[HgXe]^{2+}$ and $[AuXe_2F]^{2+}$, however, are stable at room temperature and have been fully structurally characterised as having Xe–Hg and Xe–Au bonds.

Krypton compounds

Krypton forms the highly reactive and **unstable KrF₂** by reaction of Kr with **atomic** fluorine. Compound KrF_2 decomposes to Kr and F_2 at room temperature and forms KrF^+ and $Kr_2F_3^+$ ions with fluoride ion acceptors. Bonds between krypton and elements other than fluorine are rare but can be made, generally when the subsidiary structure is resistant to oxidation.

$$3KrF_2 + 2B(OTeF_5)_3 \rightarrow 3Kr(OTeF_5)_2 + 2BF_3$$
$$KrF_2 + CF_3CN \rightarrow [CF_3CN–Kr–F]^+ [F]^-$$
$$Kr + HCN \rightarrow H–Kr–CN \text{ (low temperature photolysis)}$$
$$2Kr + O_3^+ \rightarrow KrO^+ + KrO_2^+ \text{ (observed in mass spectrometer)}$$

Compounds of radon and argon

Several radon compounds are thought to exist, but the absence of any stable isotopes (^{222}Rn, with $t_{\frac{1}{2}} = 3.8\,d$ being the most stable) means that only deductions by **tracer experiments** can be made. Structural and spectroscopic studies are not possible because of the very small amounts of material available. Evidence suggests species such as RnF^+ can be formed. In the year 2000, the first compound of argon was detected by infrared spectroscopy; HArF is stable at 10K but decomposes when warmed above 27K.

4.13. Some Industrial Processes Involving *p*-Block Elements (I)

Aluminium production

Aluminium comprises about 8% of the Earth's crust and is the third most abundant terrestrial element. It has widespread uses from packaging to aircraft construction. The metal is electropositive, hence its compounds are difficult to reduce by chemical means. The element must thus be prepared by **electrolysis**. Electrolysis in aqueous solution is not feasible as the metal reacts with water as soon as it forms, as the reduction potentials below illustrate:

$$Al^{3+} + 3e^- \rightarrow Al \qquad E^0 = -1.66 \, V$$
$$\text{hence} \quad 2Al + 6H^+ \rightarrow 2Al^{3+} + 3H_2 \qquad E^0_{cell} = 1.66 \, V$$

For this reason the electrolysis has to be carried out on molten salts. The commonest ore of aluminium is **bauxite** which has a composition ranging between $AlO(OH)$ and $Al(OH)_3$ depending on its source. The bauxite is purified by 'dissolving' in strong NaOH solution to form aluminates; the impurities are not soluble. Aluminium in the form of the hydroxide is then precipitated by reaction with $CO_2(g)$.

$$\text{bauxite} + OH^- \rightarrow [Al(OH)_4]^-$$
$$[Al(OH)_4]^- + CO_2 \rightarrow Al(OH)_3 + HCO_3^-$$
$$\text{heat}$$
$$2Al(OH)_3 \rightarrow Al_2O_3 + 3H_2O$$

Aluminium oxide has a very high melting point (over 2000°C) and is mixed with Na_3AlF_6 (cryolite) to lower this to under 1000°C and hence reduce energy costs. Electrolysis of this melt produces metallic aluminium at the cathode.

Alkene polymerisation

Organoaluminium compounds such as triethylaluminium, Et_3Al, react with alkenes to give a mixture of higher aluminium alkyls as indicated in Figure 4.13a. Triethylaluminium exists in equilibrium with its dimer, Al_2Et_6. The monomeric form, having a vacant *p*-orbital, can accept electron density from the π-bond of the alkene. One of the alkyl groups bonded to the Al then migrates to the alkene forming a C–C bond and increasing the chain length. In this type of reaction chain lengths in the range C_{14}–C_{20} are obtained. Controlled oxidation of the alkyls gives alkoxides, $Al(OR)_3$, which are then hydrolysed with acid to give a mixture of alcohols. These are used as the feedstock for the production of **biodegradable detergents**.

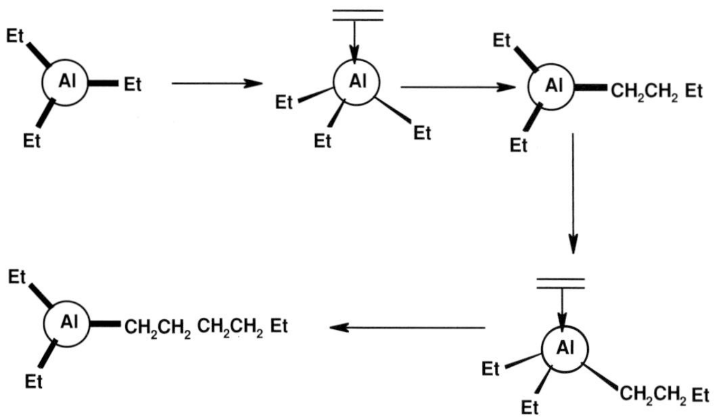

Figure 4.13a   The mechanism of alkene polymerisation by triethylaluminium.

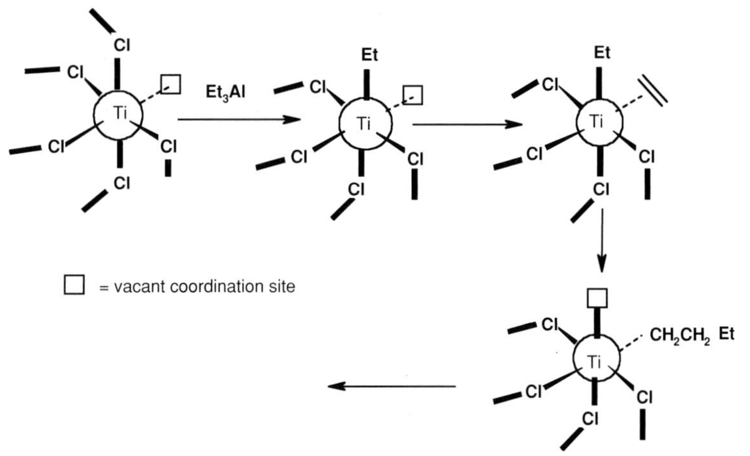

Figure 4.13b The mechanism of Ziegler–Natta catalysis. The exact nature of the catalyst is unknown. It is thought to be a polymeric solid which contains Ti(+3) at its active sites. The chlorides act as bridges to the other Ti centres.

The reaction of Et_3Al and $TiCl_4$ in hydrocarbon solvents generates a suspension containing the catalytically active, partially alkylated $TiCl_3$. This is a highly effective alkene polymerisation catalyst and gives much higher degrees of polymerisation. This type of catalyst, and many variants on it, are named **Zeigler–Natta** catalysts after their discoverers. These catalysts are used in the industrial production of high-density polythene and polypropylene. The mechanism of the reaction is very similar to that of the Et_3Al–alkene reaction in that a vacant site on the metal, this time a *d*-orbital, is available for alkene co-ordination as shown in Figure 4.13b. Not only does the catalyst give high molecular weight polymers but these are also **stereo-regular** and the polymers have more useful physical properties. The main products are high-density polythene and polypropylene. In the case of polypropylene the methyl groups are arranged on the same side of the polymer chain, as illustrated in Figure 4.13c.

Bromine

Bromine is a volatile (boiling point 59°C), toxic and reactive liquid which is produced by oxidation of bromide (often from seawater) with chlorine:

$$2Br^- + Cl_2 \rightarrow Br_2 + 2Cl^-$$

The subsequent isolation of bromine provides an excellent example of the variation of the stability of oxidation states with pH. The bromine initially formed is evaporated from solution on a current of air and concentrated by dissolving in alkaline solution where it **disproportionates** to bromide and bromate:

$$3Br_2 + 3CO_3^{2-} \rightarrow 5Br^- + BrO_3^- + 3CO_2$$

As can be seen from the **Latimer diagram** (Section 1.5) in Figure 4.13d, bromine is thermodynamically unstable with respect to **disproportionation at pH 10**. On lowering the pH of the solution to pH 0 it can be seen that the element is stable in aqueous solution, and this is used to liberate the element from the concentrated bromide/bromate solution as the reverse reaction (a **comproportionation**) becomes favourable:

$$BrO_3^- + 5Br^- \rightarrow 3Br_2 \qquad E_{cell}^0 = 0.44\,V$$

Once used to produce additives for petrol the increased use of unleaded fuels has now made this application negligible. Other industrial uses of bromine continue to expand and it finds application in the production of flame-retardant polymers, pharmaceuticals, pesticides and disinfectants.

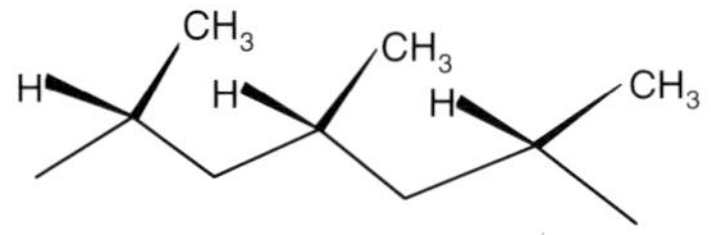

Figure 4.13c The structure of isotactic polypropylene – the methyl groups are on the same side of the polymer chain.

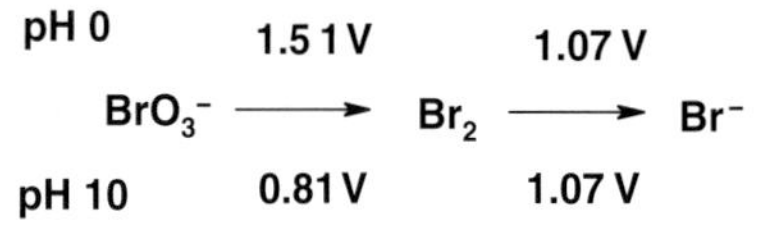

Figure 4.13d The Latimer diagram for bromine in acidic and alkaline solutions.

4.14. Some Industrial Processes Involving *p*-Block Elements (II)

Ammonia

With an annual production in the order of 10^{11} tonnes, more moles of **ammonia** are manufactured than any other chemical. Modern production of ammonia is by the iron-catalysed reaction of N_2 with H_2 and is based on the original **Haber process**, first used in 1913.

$$N_2(g) + 3H_2(g) \rightarrow 2NH_3(g) \qquad \Delta H_f^0 \; NH_3 = -46.1 \; kJmol^{-1}$$

The direct combination of nitrogen with hydrogen is exothermic (see above) but the reaction is exceedingly **slow at room temperature** even in the presence of a suitable catalyst. The obvious solution to this problem is to increase the rate of reaction by raising the temperature. Since the reaction leads to a reduction in the number of moles of gas (from 4 to 2) we have a **decrease in entropy** ($\Delta S^0 = -99.4 \, JK^{-1} \, mol^{-1}$ in this case). The effect of changing the temperature on the extent of this reaction can be approximated by substitution of the values of ΔH^0 and ΔS^0 into the equation $\Delta G = \Delta H^0 - T\Delta S^0$, where ΔG is the change in Gibbs free energy. A lowering of free energy means the process is favourable, i.e. the sign of ΔG must be negative and the greater its magnitude the more favourable the process, and in this case, the higher the equilibrium concentration of ammonia. For the reaction above we get $\Delta G = -46.1 + 0.1T$, and thus on raising the temperature the reaction becomes **less favourable** even though it will proceed at a faster rate. In cases such as this there is an optimum set of conditions which gives a reasonable yield of ammonia at an acceptable rate. Such conditions are found by trial and error and for a finely divided iron catalyst temperatures of around 400°C and pressures of around 200 atmospheres are used.

The reason for the slow reaction is the strong N≡N triple bond which must be broken at some stage in the formation of ammonia. Although the catalyst reduces the activation energy for the process, this step is still the slowest in the ammonia synthesis. The mechanism has been studied in detail and is indicated in Figure 4.14a. The reaction occurs by the **adsorption** (binding of a substance to the surface of a material) of nitrogen and hydrogen on the iron catalyst (step 1). Here the molecules are weakly bound, **physically adsorbed**, and retain their chemical identity. A stronger interaction then occurs at the surface leading to the **chemical adsorption** of the N_2 and H_2 which dissociate into **atoms** (step 3). These are much more reactive and can combine to form ammonia on the surface (steps 4 and 5) which is desorbed to expose fresh surface for further reaction. Experimental verification of this mechanism comes from **isotope redistribution** reactions in which labelled nitrogen $^{15}N_2$ and normal $^{14}N_2$ are mixed. In the absence of a catalyst no redistribution of the label occurs and only $^{15}N_2$ and $^{14}N_2$ are detected. In the presence of the catalyst, however, $^{15}N^{14}N$ is formed; the simplest way that this can be explained is by the formation of atomic species at the catalyst surface.

Figure 4.14a A mechanism for the formation of ammonia at an iron surface.

Nitric acid

One of the major uses of the ammonia produced above is in the manufacture of nitric acid. Early production of nitric acid was by the reaction of nitrates with concentrated sulfuric acid, an expensive and difficult operation to carry out on a large scale.

$$2KNO_3 + H_2SO_4 \rightarrow 2HNO_3 + K_2SO_4$$

With the availability of cheap ammonia other methods were sought. The most commercially attractive is the controlled oxidation of ammonia with air. The reaction requires elevated temperature and pressure (850°C at 5 atmospheres), and a platinum/rhodium catalyst is required in order for the reaction to proceed at an acceptable rate. The overall reaction is:

$$NH_3 + 2O_2 \rightarrow HNO_3 + H_2O$$

The reaction proceeds by a number of steps. The first step is the selective oxidation of NH_3 to nitric oxide, NO. Although the formation of other nitrogen oxides such as NO_2, N_2O, N_2O_4 and N_2O_5 is more thermodynamically favourable, the Pt/Rh catalyst gives NO in preference. This is an example of **kinetic control** of a reaction, where a **thermodynamically less favoured** product is formed at a **faster rate**. Following the formation of NO, further oxidation to NO_2 and dissolution in water gives nitric acid, as outlined in the equations below.

$$4NH_3 + 5O_2 \rightarrow 4NO + 6H_2O$$
$$2NO + O_2 \rightarrow 2NO_2$$
$$3NO_2 + H_2O \rightarrow 2HNO_3 + NO$$

The nitric acid can be concentrated by distillation only to about 69% as it forms a constant boiling mixture (**azeotrope**) with water. To obtain the more concentrated acid, dehydration with concentrated sulfuric acid is required.

One of the main uses of ammonia and nitric acid is in the formation of agricultural fertilisers such as ammonium nitrate. Other large-scale uses are indicated in Figure 4.14b, where the plastics industry uses HCN and urea, $(NH_2)_2CO$. Ammonium salts and urea are used as fertilisers and cyclohexanone is used to produce nylon. Nitric acid also finds use in the production of explosives.

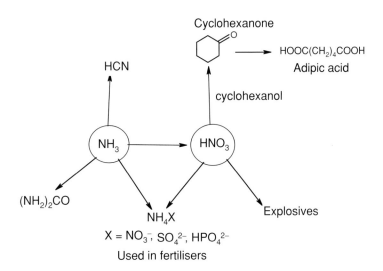

Figure 4.14b Some industrial uses of ammonia and nitric acid.

Hydrogen

5.1. Hydrogen – the First Element

Occurence

Hydrogen is the most abundant element in the universe and it is the major constituent of stars. All the other elements are believed to be obtained from it via nuclear fusion processes which are known to occur within these stars. On Earth, it is less abundant but the Earth's crust and oceans comprise $\sim 1\%$ hydrogen by weight (the ninth most abundant element). It is almost exclusively found on earth combined with oxygen as H_2O, or with carbon (and other elements) in multi-component organic based hydrocarbon deposits.

Atomic structure and abundance

The hydrogen atom has the simplest atomic structure. By definition, it has one positively charged proton in its nucleus surrounded by one electron, which in its ground state occupies the $1s$-orbital. There are five known isotopes of hydrogen (^n_1H with $n = 1–5$) with the lighter three ($n = 1, 2$ or 3) having natural terrestrial abundances of 99.98%, 0.02% and $\sim 10^{-16}\%$, respectively. The heavier nuclides ($n = 4, 5$) have been produced artificially but have short half-lives. The nuclide ^2_1H, which possess one neutron in addition to the one proton in the nucleus, is commonly referred to as deuterium and is sometimes given the symbol D. Similarly, the nuclide ^3_1H, with two neutrons and one proton in the nucleus, is commonly referred to as tritium and is sometimes given the symbol T. The lighter two isotopes ($n = 1, 2$) are stable. Tritium is radioactive, decaying by β^- emission to ^3_2He, with a half-life of 12.35 years. Tritium atoms are constantly replaced in the upper atmosphere by cosmic-ray initiated nuclear reactions.

Physical properties of the element

Terrestrially, elemental hydrogen is usually obtained under ambient conditions as a diatomic molecule, H_2. At high temperatures (e.g. 5000°C) it exists as essentially monoatomic H atoms. Physical properties of H_2, D_2 and T_2 are summarised in Figure 5.1a. The very low melting and boiling points of H_2 indicate that intermolecular forces between H_2 molecules in the solid-state and liquid-state are very weak.

	H_2	D_2	T_2
Melting point/K	14.0	18.7	20.6
Boiling point/K	20.4	23.7	25.0
Internuclear distance/pm	74.1	74.1	74.1
Bond energy/kJ mol^{-1} (at 298K)	435.9	443.4	446.9
Heat of fusion/kJ mol^{-1}	0.117	0.197	0.250
Heat of vaporisation/kJ mol^{-1}	0.904	1.226	1.393

Figure 5.1a   Some physical properties of H_2, D_2 and T_2.

Hydrogen as an industrial chemical

Hydrogen is a very important industrial chemical and it is usually obtained by **steam reformation** of natural gas at 750°C using a nickel-based catalyst. The reaction is reversible and formation of H_2 is endothermic.

$$CH_4 + H_2O \rightleftharpoons CO + 3H_2 \qquad \Delta H = +206 \text{ kJmol}^{-1}$$

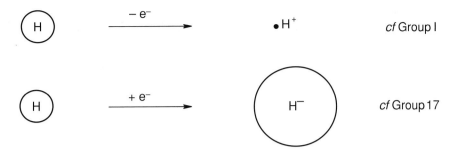

Figure 5.1b   Hydrogen has similarities in chemical properties with Group 1 and Group 17 elements.

Position in the Periodic Table

Hydrogen is unique in that it is the lightest element in the Periodic Table and the only element with one valence orbital ($1s$) and a single electron. As such, it cannot be placed unambiguously in any of the Groups of the Periodic Table.

Since it has one electron in its valence shell there are obvious similarities with the alkali metals (Group 1) with an outer configuration [inert gas core] ns^1 (Figure 5.1b). Furthermore, as would be expected from periodic trends, its (first) ionisation energy (1309 kJmol^{-1}) is higher than that of Li. However, unlike the other members of this Group, which are metallic solids under ambient conditions, it is a diatomic gas under ambient conditions.

The halogens (Group 17) are also diatomic in their elemental form, and in their atomic form are similar to hydrogen, in that each have one electron fewer in their valence shells than that required for an inert gas configuration (Figure 5.1b). For hydrogen, the closest inert gas is He ($1s^2$), the only inert gas without an outer np^6 configuration. Hydrogen is able to form a hydride anion analogous to the halide anions. In spite of this observation, the electron affinity (-79 kJmol^{-1}) and electronegativity of hydrogen are far too low for it to be placed at the head of this group. It was noted in Section 3.1 that the sodide anion, Na$^-$, with a [Ne] $3s^2$ electronic configuration, has some (limited) stability.

The electronegativity of H is too high for it to be a member of the alkali-metal family. It is more comparable to that of B (Group 13) or C (Group 14) and more typical of an element that forms covalent bonds to most elements, consistent with a partly filled valence shell.

> In conclusion, the chemical evidence (see next section) suggests that hydrogen (along with He), should be considered separately from the other elements as part of a unique introductory period ($n = 1$) to the Periodic Table.

5.2. Hydrogen – Chemical Properties

Introduction

The small size of the H atom and its moderate electronegativity value (2.2, Pauling scale) would indicate a rich and diverse chemistry for this element. Hydrogen does combine with most elements with the formation of generally strong $E-H$ bonds (see Figure 5.2a) for most non-metallic p-block elements. However, at room temperature, the H_2 molecule is relatively inert. This inertness is partly due to the high strength of the $H-H$ bond itself (432 kJ mol^{-1}) and hence this stability is kinetic rather than thermodynamic in origin.

Group	13	14	15	16	17
	B	**C**	**N**	**O**	**F**
	375	414	391	463	567
		Si	**P**	**S**	**Cl**
		323	322	364	431
		Ge	**As**	**Se**	**Br**
		289	245	277	366
				Te	**I**
				241	298

Figure 5.2a Bond enthalpy values (kJ mol^{-1}) for the $E-H$ bonds of some of the non-metallic p-block elements.

Saline hydrides

The hydrogen atom requires one electron to obtain the inert gas configuration of He (and hence a full $1s$ shell), and as such is in a similar position to that of the halogens which form stable halide ($1-$) ions by gaining an electron. It is not surprising, therefore, to note that the highly electropositive alkali metals and alkaline earths (not Be) react with H_2 at elevated temperatures to form ionic hydridic species (MH and MH$_2$ respectively) in which the hydrogen atoms are anionic with a $1-$ charge. These salts are described as **saline hydrides**. The radius of the H$^-$ ion (135–154 pm) is in between that of F$^-$ (133 pm) and Cl$^-$ (185 pm). The alkali metal hydrides adopt the cubic NaCl structure, and MgH$_2$ has a tetragonal rutile structure. The other alkaline earth metal hydrides adopt orthorhombic PbCl$_2$-type structures (Figure 5.2b). These saline hydrides react strongly with Brønsted acids to liberate H$_2$, with formation of a base.

$$H^- + H_2O \rightarrow OH^- + H_2$$

Covalent hydrides

The hydrogen atom may also obtain the full-shell electronic configuration of He by formation of a single covalent bond, and the sharing of a pair of electrons. This situation often arises when H atoms are bound to p-block elements. These molecular compounds are often referred to as **covalent hydrides**. Simple covalent hydrides include CH$_4$, NH$_3$, H$_2$O, H$_2$S, HCl, and details of these and related compounds can be found in the relevant sections. Organometallic compounds often contain M$-$H bonds, e.g. [HMn(CO)$_5$], and these can also be considered as covalent hydrides (Section 6.11). The M$-$H bonds are usually polarised $^{\delta+}$M$-$H$^{\delta-}$.

Hydrogen also partakes in **covalent multi-centre bonding** in 'electron-deficient' compounds such as the boron hydrides (Section 4.4).

66

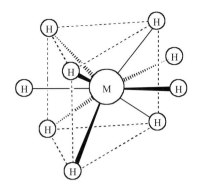

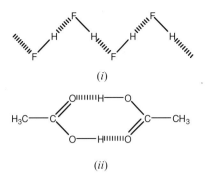

(i)

(ii)

Figure 5.2b The solid-state co-ordination geometry around the metal in MH_2 (M = Ca, Sr, Ba).

Figure 5.2c Examples of H-bonding. (i) HF, (ii) ethanoic acid.

Hydrogen bonds

Hydrogen atoms covalently bound to electronegative atoms (usually F, O or N) often form additional intermolecular or intramolecular interactions with other electronegative atoms. These additional interactions are electrostatic in origin, and are described as **hydrogen bonds** or H-bonds. The covalent H$-$E (E = electronegative element) bond is polarised $^{\delta+}$H$-$E$^{\delta-}$, and the (positively) charged $^{\delta+}$H atom is attracted to a lone-pair of electrons of a neighbouring electronegative atom. This very important interaction has a profound influence on physical properties of substances, e.g. H_2O has intermolecular H-bonds and is liquid at room temperature, whereas H-bonding is largely absent in H_2S which is a gas at room temperature (see Section 4.8). Hydrogen bonding is also of great importance in biological systems where it often controls the conformations and configurations of many biologically important molecules. Some typical examples of H-bonding are shown in Figure 5.2c and Figure 4.8c. H-bond interactions are in the range 4–40 kJ mol^{-1} and as such are typically $\sim$10% of the strength of covalent bonds.

Brønsted acidity and H$^+$

Brønsted acids are defined as compounds which are able to ionise and furnish H$^+$ ions in aqueous solution, i.e proton donors; typical examples are H_2SO_4, H_3PO_4 and HCl. In practice, the H$^+$ ion is so small and so highly Lewis acidic that it is strongly hydrated, and is better represented as the H_3O^+(aq) ion, or by the even larger $[H(OH_2)_n]^+$ (aq) (n = 1–4, 6) ions, e.g. $H_9O_4^+$(aq) (Figure 5.2d), which have additional H_2O molecules H-bonded to the H_3O^+. The H_3O^+ ion is called the **hydronium** (or alternatively named **oxonium**) ion.

$$H_2SO_4 \rightleftharpoons HSO_4^- + H^+$$
$$H_2SO_4 \, (aq) + H_2O \, (l) \rightleftharpoons HSO_4^- \, (aq) + H_3O^+ (aq)$$

The hydronium ion has been isolated in a non-hydrated form in the solid-state and has been characterised in $[H_3O][ClO_4]$ by single-crystal X-ray crystallography studies; it is pyramidal with three equivalent O$-$H distances (101 pm).

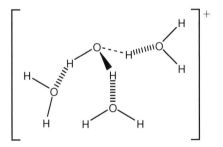

Figure 5.2d Structure of the hydronium ion, $[H_9O_4]^+$. The central O atom is pyramidal with three equivalent O$-$H bonds at 101 pm. There are three H-bonded H_2O molecules with 250–260 pm O . . . O distances.

Metallic hydrides

Many of the more electropositive d-block and f-block elements form solid **metallic hydrides** by absorbing H_2 at high temperatures. These materials have metallic appearance and properties (conduction, magnetic). The structures of these compounds are usually based on close-packed metal atom arrangements (but not necessarily those adopted by the metal itself), with H atoms occupying tetrahedral or (less frequently) octahedral holes within the lattice. Many of the compounds are non-stoichiometric and hydrogen deficient. A good illustrative example is $TiH_{1.8}$ which is idealised as MH_2. This has a fluorite structure in which the metal atoms are cubic close-packed with all the tetrahedral holes being occupied by H. Other isostructural (idealised) MH_2 species include M = Sc, Y, La, Ac, Zr, Hf, Ce, Gd, Ho, Th, Np, Pu, Am. Idealised MH_3 species have structures based on cubic (e.g. M = La, Ce, U) or hexagonal (e.g. M = Y, Gd, Ho, Np, Pu, Am) close-packed structures.

d-Block Elements – Transition Metals (Groups 3–12)

6.1. Electronic Configurations and General Reactivity

Characteristic properties of the *d*-block elements

There are several properties which are characteristic of *d*-block elements (Figure 6.1a) and these include:

> variable oxidation states,
> coloured compounds,
> unusual magnetic behaviour, and
> formation of co-ordination compounds (complexes).

All of these attributes are influenced by their electronic configurations and these important topics will be considered in more detail either here or in following sections.

Group	3	4	5	6	7	8	9	10	11	12
1st series (3*d*)	Sc	Ti	V	Cr	Mn	Fe	Co	Ni	Cu	Zn
2nd series (4*d*)	Y	Zr	Nb	Mo	Te	Ru	Rh	Pd	Ag	Cd
3rd series (5*d*)	Lu	Hf	Ta	W	Re	Os	Ir	Pt	Au	Hg
4th series (6*d*)	Lr	Rf	Db	Sg	Bh	Hs	Mt			

Figure 6.1a The *d*-block elements are a group of 37 metallic elements, numbered as Groups 3–12 and arranged in four series. The chemistry of the elements of the fourth series has not yet been investigated in detail.

Electronic configuration

The '*d*-block elements' are sometimes referred to as 'transition-elements' and the terms are almost interchangeable. However, a transition-element is one with an incomplete *d*-subshell. The *d*-block elements are arranged in **four series** (known as the 1st, 2nd, 3rd and 4th transition-series) consistent with different *d*-subshells being filled according to the aufbau principle (Figure 1.3a). The 1st transition-series relates to filling the 3*d*-subshell (after the 4*s*), and the 2nd transition-series and 3rd transition-series are concerned with filling the 4*d*- (after 5*s*) and 5*d*- (after 6*s*/4*f*) subshells, respectively. There is a 4th transition-series associated with the 6*d*-subshell (after 7*s*/5*f*), but the known chemistry of these elements is not well developed due to the lack of any stable isotopes. The ground-state gas-phase electronic configurations of the 1st transition-series in zero oxidation state are shown in Figure 6.1b. The 4*s*- and 3*d*-orbitals are of similar energy, and all these transition-elements have two electrons in the 4*s*-orbital, except for Cr and Cu which have extra stability associated with half-full and full *d*-subshell configurations. Upon oxidation the 4*s* electrons are lost in preference to 3*d* electrons.

> e.g. $Fe^{3+} = [Ar]3d^5$, $Ni^{2+} = [Ar]3d^8$, $Cu^{2+} = [Ar]3d^9$, $Cu^+ = [Ar]3d^{10}$

For the 2nd transition-series the energies of the 5*s*- and 4*d*-orbitals are very similar and several similar anomalies exist (Figure 6.1b). Similarly to Cr and Cu of the first series, Mo and Ag have a single electron in the 5*s*-subshell and half-full and full 4*d*-subshells, respectively. However, Nb, Ru and Rh also have a single electron in their 5*s*-subshell, and Pd fills its 4*d* shell and leaves its 5*s* without electrons.

70

1st transition-series (3d)		2nd transition-series (4d)		3rd transition-series (5d)	
Sc	$[Ar]4s^23d^1$	Y	$[Kr]5s^24d^1$	Lu	$[Xe]4f^{14}6s^25d^1$
Ti	$[Ar]4s^23d^2$	Zr	$[Kr]5s^24d^2$	Hf	$[Xe]4f^{14}6s^25d^2$
V	$[Ar]4s^23d^3$	Nb	$[Kr]5s^14d^4$	Ta	$[Xe]4f^{14}6s^25d^3$
Cr	$[Ar]4s^13d^5$	Mo	$[Kr]5s^14d^5$	W	$[Xe]4f^{14}6s^25d^4$
Mn	$[Ar]4s^23d^5$	Tc	$[Kr]5s^24d^5$	Re	$[Xe]4f^{14}6s^25d^5$
Fe	$[Ar]4s^23d^6$	Ru	$[Kr]5s^14d^7$	Os	$[Xe]4f^{14}6s^25d^6$
Co	$[Ar]4s^23d^7$	Rh	$[Kr]5s^14d^8$	Ir	$[Xe]4f^{14}6s^25d^7$
Ni	$[Ar]4s^23d^8$	Pd	$[Kr]5s^04d^{10}$	Pt	$[Xe]4f^{14}6s^15d^9$
Cu	$[Ar]4s^13d^{10}$	Ag	$[Kr]5s^14d^{10}$	Au	$[Xe]4f^{14}6s^15d^{10}$
Zn	$[Ar]4s^23d^{10}$	Cd	$[Kr]5s^24d^{10}$	Hg	$[Xe]4f^{14}6s^25d^{10}$

Figure 6.1b Ground-state electronic configurations of the 1st (3d), 2nd (4d) and 3rd (5d) transition-series. The underlying core inert gas configurations are $[Ar] = 1s^22s^22p^63s^23p^6$, $[Kr] = 1s^22s^22p^63s^23p^64s^23d^{10}4p^6$ and $[Xe] = 1s^22s^22p^63s^23p^64s^23d^{10}4p^65s^24d^{10}5p^6$.

The 3rd transition-series starts after completion of the lanthanide series (4f), and follows the usual pattern (Figure 6.1b): again, the 6s- and 5d-subshells are of similar energy and two electrons in the 6s orbital is the norm, except for Pt and Au, which have a single electron in that orbital.

Physical properties and general reactivity

The transition-elements are all metals with **typical metallic properties**. Most d-block elements are hard metallic solids (Section 2.3) which are ductile and malleable. However, Hg is a liquid at room temperature. All d-block elements have high thermal and electrical conductivities.

The **reactivity** of bulk d-block elements towards O_2, halogens, acids, etc., is variable and ranges from moderate (the majority) to low (late 2nd and 3rd series metals). However, finely-divided metals can be much more reactive, for example Fe and Ni are pyrophoric! Au is unreactive towards O_2 and acids and is only attacked by aqua regia (a 3:1 mixture of concentrated HCl and HNO_3 acids).

The early d-block elements are commonly found in oxide-containing **ores** (e.g. ilmenite $FeTiO_3$, scheelite $CaWO_3$), whereas the late 2nd and 3rd series elements are commonly found combined with sulfur (e.g. argentite Ag_2S, cinnabar HgS) or native (Au, Hg, Pt). The late 1st series metals are found combined with either O or S (e.g. haematite Fe_2O_3, iron pyrite FeS_2, chalcopyrite $CuFeS_2$, sphalerite ZnS). These naturally occurring ores reflect the metal's transition from 'hard' to 'soft', upon progressing across and descending the d-block. Iron is the fourth most abundant element (after O, Si and Al) in the Earth's crust.

There is a large increase in atomic size on progressing down a d-block triad between the 1st transition-series and 2nd transition-series elements, but because of the **lanthanide contraction** (Section 7.1) the 3rd transition-series metal is of a similar radius to that of the 2nd series, e.g. Cr (125 pm), Mo (136 pm), W (137 pm); and Fe (126 pm), Ru (133 pm), Os (134 pm) (Figure 6.1c). Properties that depend upon atomic size are therefore similar for the 2nd and 3rd series elements but differ from the 1st transition-series.

161	145	131	125	129	126	125	124	128	133
Sc	**Ti**	**V**	**Cr**	**Mn**	**Fe**	**Co**	**Ni**	**Cu**	**Zn**
144	132	122	117	117	116	116	115	117	125

178	159	141	136	135	133	134	138	144	149
Y	**Zr**	**Nb**	**Mo**	**Tc**	**Ru**	**Rh**	**Pd**	**Ag**	**Cd**
162	145	134	129	127	124	125	128	134	141

172	157	143	137	137	134	135	138	144	152
Lu	**Hf**	**Ta**	**W**	**Re**	**Os**	**Ir**	**Pt**	**Au**	**Hg**
160	144	134	130	128	126	126	129	134	144

Figure 6.1c Metallic and covalent radii of the d-block elements. The metallic radius (pm) is above the element symbol and the covalent radius (pm) is below.

6.2. An Overview of Chemical and Physical Properties of the *d*-Block Elements

Co-ordination compounds (formation of complexes)

The *d*-block metal ions are far too Lewis acidic ever to be found 'naked', except for, perhaps, in the gas phase. The *d*-block metals readily partake in complex formation where the metallic atom is bound by a number of donor ligands by co-ordinate links. Co-ordination compounds (or complexes) were first described by **Werner** in a publication on cobaltammines (1891). He described co-ordination compounds as those in which the transition-metal's **secondary valency** was greater than its **primary valency**. Nowadays, we refer to primary valency as **oxidation state** and secondary valency as **co-ordination number**. Common co-ordination numbers (2, 4, 6) and co-ordination geometries are reviewed in Section 6.4. Higher co-ordination numbers, with consequently different geometries, may be observed for the 1st transition-series, but are more commonly observed in compounds of the larger elements of the 2nd and 3rd transition-series (and the *f*-block metals, Section 7).

Coloured compounds

Virtually all compounds of the transition-elements are weakly coloured. In general, a substance will appear coloured if it absorbs some of the light that falls upon it. The light which is then transmitted or reflected back to the observer's eyes appears to have a colour complementary to that of the absorbed light. The absorption of visible radiation invariably involves promotion of an electron from one energy level to another, and the **partially filled *d*-orbitals** of the transition-metal are a potential cause of colour in their compounds. Compounds of the d^0 (Sc^{3+}, Ti^{4+}) and d^{10} (Cu^+, Zn^{2+}) metal ions are generally colourless. However, intense colours may also be due to **charge transfer bands** such as in MnO_4^- where the d^0 Mn(+7) ion is present. See ligand field theory (Section 6.8) and electronic spectra (Section 6.9) for more details.

Paramagnetism

A substance which is **attracted into a magnetic field** is said to be **paramagnetic** whereas if it is repelled it is said to be diamagnetic. Transition-metals and their ions are commonly paramagnetic and this behaviour contrasts with main group elements whose compounds are almost exclusively diamagnetic. Paramagnetism in transition-metals is associated with **unpaired electrons** in their partially filled *d*-orbitals. The occurrence of unpaired electrons in transition-metal complexes is often a consequence of the number, geometry and electronic properties of the ligands bound to the metal centre. See ligand field theory (Section 6.8) and magnetic properties (Section 6.10) for more details.

Catalysis involving *d*-block metals

The rate of homogeneous or heterogeneous reactions may sometimes be increased by the addition of a catalyst, which provides an alternative lower activation energy route to the same products. The catalyst is not consumed in the process. Transition-metals and their compounds are commonly found to be useful catalysts and this is often associated with their ability to exhibit **variable oxidation states**, and the ability of the metal to cycle between oxidation states. The available oxidation states of transition-metals are reviewed in Section 6.3. **Flexibility of co-ordination number/geometry** is also an important factor in catalysis and co-ordination numbers/geometries are reviewed in Section 6.4. Examples of some important industrial reactions catalysed by transition-metal catalysts are shown below:

$$2SO_2 + O_2 \rightarrow 2SO_3 \qquad \text{catalyst} = V_2O_5$$
$$N_2 + 3H_2 \rightarrow 2NH_3 \qquad \text{catalyst} = Fe$$
$$R_2C{=}CR_2 + H_2 \rightarrow R_2CHCHR_2 \qquad \text{catalyst} = Ni$$

The catalytic industrial production of SO_3 and NH_3 are described in more detail in Sections 4.10 and 4.14, respectively.

d-Block metal complexes are also of vital importance in **biological systems** where **metalloenzymes** are found to catalyse or facilitate specific processes, some familiar examples being:

> Fe: Haemoglobin, myoglobin (O_2 transport) (Figure 6.2a)
> Co: Vitamin B_{12} ('isomerase' reactions, and methyl group transfer reactions)
> Cu/Zn: Superoxide dismutase (decomposition of O_2^-) (Figure 6.2b)
> Zn: Carbonic anhydrase (CO_2 transport)

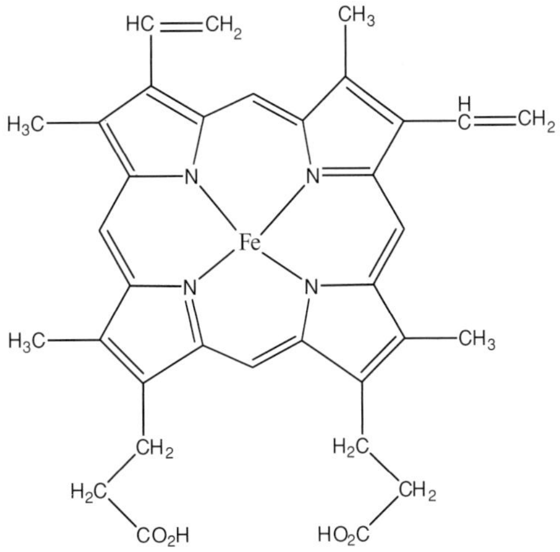

Figure 6.2a The Fe(+2) complex of haem-b plays an important role in O_2 transport.

The human body typically contains ~5 g Fe, ~2 g Zn and ~0.1 g Cu. Other *d*-block metals are found in much smaller quantities e.g. Co ~0.003 g. *d*-Block metals which are known to be essential to human beings, and hence must play important biological roles are V, Cr, Mo, Mn, Fe, Co, Ni, Cu, Zn. Manganese (along with Mg) is involved in photosynthesis in plants. Molybdenum and Fe play important roles in nitrogenases (Figure 6.2c). Nitrogenases are the metalloenzymes responsible for 'biological fixation of nitrogen' whereby some bacteria and blue-green algae convert atmospheric N_2 into NH_3.

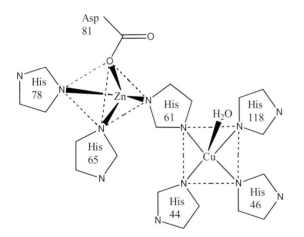

Figure 6.2b The active site of bovine superoxide dismutase contains a Cu(II) and a Zn(II) centre.

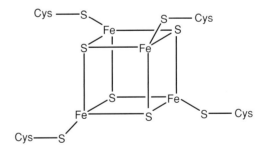

Figure 6.2c Two protein components, the Fe protein and Fe–Mo protein, make up the nitrogenase enzyme complex. In the Fe-protein the Fe is present mainly as cubane-like Fe_4S_4 clusters and these clusters are involved in electron storage/transfer.

6.3. Variable Oxidation States

The first transition-series (3*d*)

The similarity in energy of the 4*s*- and 3*d*-orbitals leads to the observation of **variable oxidation states** for first series transition-metals. This is shown schematically in Figure 6.3a. Qualitatively, four general points to note for this series are:

(1) All elements (except Zn) are able to form complexes in the +3 oxidation state.
(2) The +2 oxidation state becomes increasingly more stable on going from left to right, with Ti^{2+} through Cr^{2+} being reducing agents and Mn^{2+} through Cu^{2+} stable, with Cu^{2+} weakly oxidising.
(3) The maximum oxidation state for Sc through Mn is equal to the Group oxidation number (i.e. 3 to 7). This corresponds to the total number of valence electrons (4*s* + 3*d*) available to the metal.
(4) For Groups 3, 4 and 5 the Group oxidation state is stable but Cr(VI) (Group 6) and Mn(VII) (Group 7) are oxidising.

Quantitative free energy oxidation state diagrams (Section 1.6) for some elements of the first transition-series are also shown in Figure 6.3b.

The oxidation states, available in aqueous solution, of V, Cr and Mn, are described in more detail below. All ions in aqueous solution are hydrated although this may not be explicit in their formulae.

Sc			3				
Ti		2r*	3r	4			
V		2r*	3r	4	5		
Cr		2r	3	4d*	5d*	6o	
Mn		2	3o*	4o	5d*	6d*	7o
Fe		2	3	4o*	5o*	6o*	
Co	1r*	2	3	4o*			
Ni	1r*	2	3o*				
Cu	1	2	3o*				
Zn		2					

Figure 6.3a Oxidation states of the 3*d* transition-elements as observed in their oxides, halides, and in aqueous solution. Oxidation states in bold are the most stable, and those marked* are the least stable (r = reducing, o = oxidising, d = disproportionates).

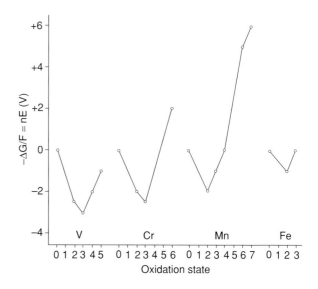

Figure 6.3b Free energy oxidation state (FROST) diagrams for selected 3*d* elements: V, Cr, Mn and Fe.

Group 5 – oxidation states of V

The highest oxidation state is +5. This exists in acid solution as the orange coloured dioxovanadium(V) cation, VO_2^+, and in alkaline solution as a colourless VO_4^{3-} anion. The VO_2^+ ion is angular (cis) rather than linear (trans). It can be reduced by Zn/HCl to the blue coloured ***vanadyl cation*** VO^{2+}, which has the vanadium in the +4 oxidation state. There is an extensive co-ordination chemistry associated with these dioxovanadium(V) and oxovanadium(IV) cations. Chemical or electrolytic reduction of the vanadyl cation leads to the relatively stable green solutions containing V^{3+}, and further reaction ultimately yields the strongly reducing lilac coloured V^{2+}. The V^{2+} cation is readily oxidised by air (O_2), and slowly by H_2O, to V^{3+}.

74

Group 6 – oxidation states of Cr

The highest oxidation state is +6. In acidic solutions it exists mainly as the orange dichromate ($Cr_2O_7^{2-}$) anion whilst in alkaline solutions it exists as the yellow chromate(VI) anion, CrO_4^{2-}.

$$2H^+ + 2CrO_4^{2-} \rightleftharpoons Cr_2O_7^{2-} + H_2O$$

The CrO_4^{2-} anion is tetrahedral at Cr, and the $Cr_2O_7^{2-}$ anion has the structure of two tetrahedral Cr centres bridged by a shared oxygen atom. Dichromate, under acid conditions, is a well-known oxidising agent, whereas in alkaline conditions Cr(VI) is only weakly oxidising:

$$Cr_2O_7^{2-} + 14H^+ + 6e^- \rightleftharpoons 2Cr^{3+} + 7H_2O \qquad E^0 = 1.33\,V$$
$$CrO_4^{2-} + 4H_2O + 3e^- \rightleftharpoons Cr(OH)_3 + 5OH^- \qquad E^0 = 0.13\,V$$

Anion $Cr_2O_7^{2-}$ can be reduced by Zn/HCl to the green coloured and stable Cr(III) species, $[CrCl(H_2O)_5]^{2+}$. The Cr^{3+}(aq) cation is violet (see Section 6.6). Further reduction with Zn/HCl produces the blue coloured strongly reducing Cr^{2+} which is readily oxidised by air (O_2) back to Cr^{3+}.

Group 7 – oxidation states of Mn

The highest oxidation state (+7) is observed in the purple coloured, strongly oxidising permanganate MnO_4^- anion. It can be reduced by an excess of various reducing agents to the Mn^{2+} cation which is very pale pink coloured and stable to air (O_2). However, if the MnO_4^- is in excess it is able to re-oxidise the Mn^{2+} to the black coloured MnO_2.

$$MnO_4^- + 8H^+ + 5e^- \rightleftharpoons Mn^{2+} + 4H_2O \qquad E^0 = 1.52\,V$$
$$2MnO_4^- + 3Mn^{2+} + 2H_2O \rightleftharpoons 5MnO_2 + 4H^+ \qquad E^0_{cell} = 0.46\,V$$
$$MnO_4^- + 2H_2O + 3e^- \rightleftharpoons MnO_2 + 4OH^- \qquad E^0 = 1.23\,V$$

Under basic conditions MnO_4^- is also a strong oxidant, and may be reduced to MnO_2. The deep green manganate(VI) ion MnO_4^{2-}, is stable in alkaline solution but disproportionates to MnO_4^- and Mn^{2+} in neutral or acid conditions.

Oxidation states of the second (4d) and third (5d) transition-metal series

The **oxidation states** of these elements follow a similar pattern to those observed for the first series, with higher oxidation states generally more stable than lower oxidation states (+2, +3) (Figure 6.3c). The group oxidation state is available as far to the right of the series as Group 8 (e.g. RuO_4, OsO_4), and higher oxidation states than those commonly observed in the 3d series are observed in the later elements of the second and third series. With the exception of Y^{3+}(aq) simple aqueous cations are not commonly observed for elements of the second and third transition-series, and high oxidation state oxoanions, e.g. TcO_4^-, ReO_4^-, are less strongly oxidising.

Y		3						
Zr	2r	3r*	4					
Nb	2r	3r	4d*	5				
Mo		3	4	5	6			
Tc			4	5d*	6	7		
Ru		2	3	4	5*	6	7d*	8o
Rh 1*		2	3	4		6o		
Pd		2		4				
Ag 1		2o	3o*					
Cd		2						

Lu		3						
Hf	2r*	3r*	4					
Ta	2*	3r	4d*	5				
W	2*	3	4	5	6			
Re 1*	2*	3	4	5*	6	7		
Os	2*	3	4	5*	6	7*	8o	
Ir 1		3	4	5o*	6o*			
Pt	2		4	5o*	6o*			
Au 1		3						
Hg 1	2							

Figure 6.3c Oxidation states of the 4d and 5d transition-elements as observed in their oxides, halides, and in aqueous chemistry. Oxidation states in bold are the most stable, and those marked * are the least stable (r = reducing, o = oxidising, d = disproportionates).

6.4. Co-ordination Compounds (I)

Ligands and denticity

Ligands are anions or neutral molecules which can be thought of as electron pair donors or Lewis bases. The number of simultaneous 2-electron donor bonds that a ligand can make to the Lewis acidic metal ion is described as its **denticity**.

Examples of **monodentate** ligands:	F^-, Cl^-, Br^-, I^-, CN^-, OH^-, H_2O, NH_3, CO, MeOH.
Examples of **bidentate** ligands:	$H_2NCH_2CH_2NH_2$ (en, ethylenediamine)
	$[O_2CCO_2]^{2-}$ (ox, oxalate)

Tridentate, tetradentate, etc., ligands also exist and these are described as **polydentate** ligands. Examples of polydentate ligands are shown in Figure 3.1b and Figure 6.4a.

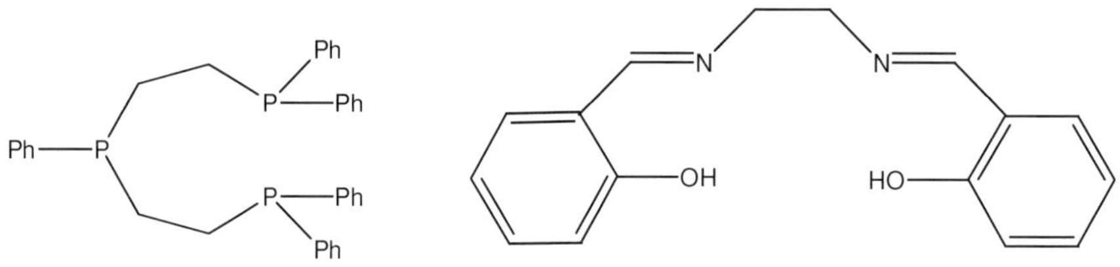

Figure 6.4a Examples of (i) a tridentate and (ii) a tetradentate ligand.

Co-ordination numbers and co-ordination geometries

The **co-ordination number** is the number of groups that immediately surround the metal. It is important to know the spatial arrangements of these groups. This is described as the **co-ordination geometry** (Figure 6.4b). There is a definite correspondence between the two but the situation is more complicated than VSEPR (Section 2.2), as used in main-group chemistry, due to the presence of the metal's d-electrons.

A **co-ordination number of 2** is generally **rare** but it is observed in the +1 cations of the coinage metals, (Cu^+, Ag^+, Au^+). The angles between ligands surrounding the metal centre are $180°$ i.e. a **linear geometry**. These complexes may add a further two ligands and become 4 co-ordinate.

2 co-ordinate:	$[Ag(NH_3)_2]^+$, $[CuCl_2]^-$, $[Cu(CN)_2]^-$.
4 co-ordinate:	e.g. $[Cu(CN)_2]^- + 2CN^- \rightarrow [Cu(CN)_4]^{3-}$

A **co-ordination number of 4** is **fairly common** in transition-metal chemistry. Such complexes e.g. $[Cu(CN)_4]^{3-}$ are usually **tetrahedral**, with angles between ligands of $109.5°$. However, metal ions with d^8 **configuration** [Ni(II), Pd(II), Pt(II), Rh(I), Ir(I), Au(III)] and a co-ordination of 4 are generally **square-planar** e.g. cis-$[PtCl_2(NH_3)_2]$. This geometry change is a direct consequence of the d-electrons stabilising this particular geometry (Section 6.8). The square-planar geometry has interligand angles of $90°$. It is also observed for Cu(II) (d^9 configuration) in some of its complexes.

A **co-ordination number of 6** is **very common** in transition-metal complexes where an **octahedral** geometry is adopted. Most M^{2+} and M^{3+} transition-metal ions form octahedral complexes in aqueous solution. All $M-L$ bonds are equivalent in an octahedron and angles between ligands are $90°$ or $180°$.

e.g. $Mn^{2+}(aq) = [Mn(H_2O)_6]^{2+}$

Sometimes the H_2O ligands are omitted from the formula for clarity.

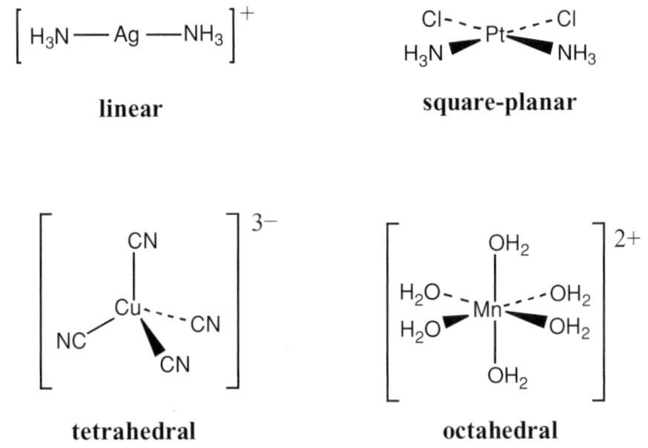

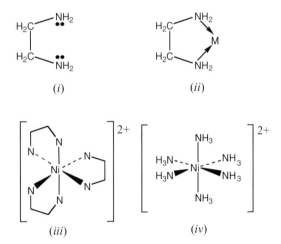

Figure 6.4b Examples of complexes in commonly observed co-ordination numbers/geometries.

The chelate effect

Bidentate and higher denticity ligands often form ring structures upon co-ordination to metal ions (Figure 6.4c). This is termed **chelation** (from the Greek, for pincer, claw). The formation of 5-membered rings leads to particularly stable complexes. Complexes with chelate rings are generally much more (thermodynamically) stable than related complexes that contain an equivalent number of monodentate ligands. A good illustration is to compare $[Ni(NH_3)_6]^{2+}$ and $[Ni(en)_3]^{2+}$ which have overall stability constants of $10^{8.6}$ and $10^{18.5}$ for β_6 and β_3, respectively. The chelate effect originates primarily from **entropy** (ΔS) rather than enthalpy (ΔH) changes.

Figure 6.4c The chelate effect. (i) Ethylenediamine (1,2-diaminoethane, or en) is a potentially bidentate ligand, binding through both N atoms. (ii) Upon co-ordination a five-membered ring is formed. (iii) The complex $[Ni(en)_3]^{2+}$, is more stable than the equivalent acyclic complex (iv) $[Ni(NH_3)_6]^{2+}$. This extra stability originates, in part, from an entropy effect since there is a net increase in the number of particles $(7 - 4 = 3)$ upon formation of $[Ni(en)_3]^{2+}$ (and $6H_2O$) from $[Ni(H_2O)_6]^{2+}$ (and 3en).

6.5. Co-ordination Compounds (II)

Kinetically inert and labile complexes

For most transition-metals substitution or ligand-exchange reactions are extremely fast, and complexes that undergo **fast ligand exchange** are termed **labile**. Ligand field stabilisation of some transition-metal ions {Cr^{3+} (d^3), Fe^{2+} (d^6), Co^{3+} (d^6), and d^6 and d^8 ions of the $4d$ and $5d$ series} may **slow ligand-exchange** reactions by several orders of magnitude, and such stabilised complexes are termed **kinetically inert**. However, since most complexes are labile and reaction equilibria are attained rapidly, the position of equilibrium is determined by the thermodynamics of the system and described by equilibrium (stability) constants.

Stepwise and overall stability constants

Two related types of stability constants are used to describe the thermodynamic stabilities of complexes in solution. These are **stepwise stability** or **stepwise formation constants (K_n)**, and the **overall stability** or **overall formation constants (β_n)**. These relate to the following equilibria:

Stepwise:

$M + L \rightleftharpoons ML \qquad K_1 = [ML]/[M][L]$

$ML + L \rightleftharpoons ML_2 \qquad K_2 = [ML_2]/[ML][L]$

$ML_2 + L \rightleftharpoons ML_3 \qquad K_3 = [ML_3]/[ML_2][L]$

General case with n = maximum co-ordination number of the metal:

$ML_{n-1} + L \rightleftharpoons ML_n \qquad K_n = [ML_n]/[ML_{n-1}][L]$

Overall:

$M + L \rightleftharpoons ML \qquad \beta_1 = [ML]/[M][L]$

$M + 2L \rightleftharpoons ML_2 \qquad \beta_2 = [ML_2]/[M][L]^2$

$M + 3L \; ML_3 \qquad \beta_3 = [ML_3]/[M][L]^3$

General case with n = maximum co-ordination number of the metal:

$M + nL \rightleftharpoons ML_n \qquad \beta_n = [ML_n]/[M][L]^n$

Stepwise formation constants can be used to determine the exact concentrations of all the various complexes at equilibrium at defined metal and ligand concentrations. Overall stability constants describe the thermodynamic stability of the complex. Stepwise and overall stability constants are related to one another since rearrangement of these equations leads to $\beta_n = K_1.K_2.K_3...K_n$. Examples of overall and stepwise equilibrium constants are given in Figure 6.5a.

Equilibrium				$\log\beta_n$
Ag^+	+	$2NH_3$	$\rightleftharpoons$ $[Ag(NH_3)_2]^+$	7.2
Ag^+	+	$2S_2O_3^{2-}$	$\rightleftharpoons$ $[Ag(S_2O_3)_2]^{3-}$	13.0
Co^{2+}	+	$6NH_3$	$\rightleftharpoons$ $[Co(NH_3)_6]^{2+}$	4.9
Co^{3+}	+	$6NH_3$	$\rightleftharpoons$ $[Co(NH_3)_6]^{3+}$	34.4
Cu^+	+	$4CN^-$	$\rightleftharpoons$ $[Cu(CN)_4]^{3-}$	27
Cu^+	+	$2NH_3$	$\rightleftharpoons$ $[Cu(NH_3)_2]^+$	11
Cu^{2+}	+	$4Cl^-$	$\rightleftharpoons$ $[CuCl_4]^{2-}$	5.6
Cu^{2+}	+	$4NH_3$	$\rightleftharpoons$ $[Cu(NH_3)_4]^{2+}$	13.1
Fe^{2+}	+	$6CN^-$	$\rightleftharpoons$ $[Fe(CN)_6]^{4-}$	35
Fe^{3+}	+	$6CN^-$	$\rightleftharpoons$ $[Fe(CN)_6]^{3-}$	43
Fe^{3+}	+	$4Cl^-$	$\rightleftharpoons$ $[FeCl_4]^-$	−1.1
Ni^{2+}	+	$6NH_3$	$\rightleftharpoons$ $[Ni(NH_3)_6]^{2+}$	8.3
Zn^{2+}	+	$4NH_3$	$\rightleftharpoons$ $[Zn(NH_3)_4]^{2+}$	9.6

Stepwise formation constants (K_n) for Cu^{2+}/NH_3:
$\log K_1$ 4.3; $\log K_2$ 3.6, $\log K_3$ 3.0, $\log K_4$ 2.2; ($\beta_4 = K_1.K_2.K_3.K_4$)

Figure 6.5a Stability constants for some transition-metal complexes.

Irving–Williams series for M^{2+}

The relative stability of complexes with divalent metals generally follows the order:

$$Ba^{2+} < Sr^{2+} < Ca^{2+} < Mg^{2+} < Mn^{2+} < Fe^{2+} < Co^{2+} < Ni^{2+} < Cu^{2+} > Zn^{2+}$$

This order is known as the **Irving–Williams series** of stability and is explained as being electrostatic in origin since the order is the **reverse of cation size**. This is illustrated by $M(NH_3)_4^{2+}(aq)$ complexes where the following $\log\beta_4$ values have been obtained: Co^{2+} (5.1) $< Ni^{2+}$ (7.7) $< Cu^{2+}$(13.1) $> Zn^{2+}$(9.6).

The effect of metal charge on stability constants

Increasing the charge on the metal centre whilst keeping the ligands surrounding the metal constant, leads to **higher stability constants** as exemplified by $\log\beta_6$ values for $[Co(NH_3)_6]^{2+}$ (4.9) and $[Co(NH_3)_6]^{3+}$ (34.4). This is an electrostatic effect.

Substitution in square-planar Pt(II) complexes

Square-planar complexes of low-spin d^8 ions, e.g. Pt^{2+}, are kinetically inert. The rate of ligand exchange is dependent upon the ease of formation of a 5 co-ordinate (trigonal bipyramidal) intermediate. For Pt^{2+} the rate of reaction is increased by increasing the polarisability of both the incoming and the leaving groups, and by the electronic properties of the ligand *trans* to the leaving group. The latter is termed the **trans effect** and a 'trans effect scale' may be set up which indicates the ability of a ligand to labilise the ligand *trans* to it.

$$H_2O, NH_3 < Cl^- < Br^- < H^- < CO, CN^-$$

The trans effect often dictates the stereochemistry observed in synthetic transformations (Figure 6.5b). It is a kinetic phenomenon influenced by both ground state destabilisation (thermodynamic) and transition state stabilisation.

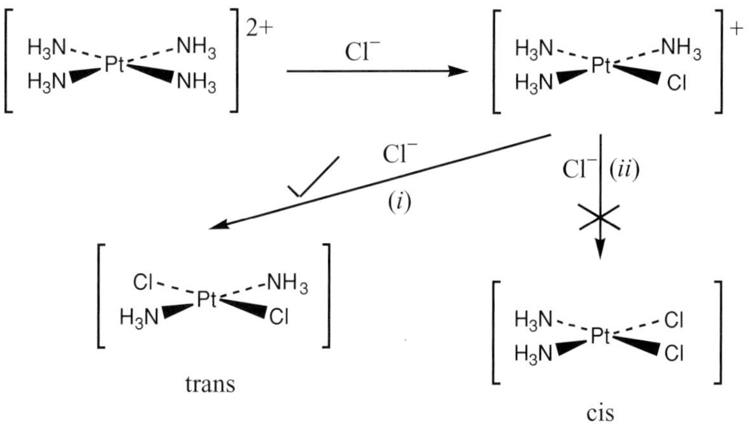

Figure 6.5b   Treatment of $[Pt(NH_3)_4]^{2+}$ with two equivalents of Cl^- results in the *trans*-$[PtCl_2(NH_3)_2]$ rather than *cis*-$[PtCl_2(NH_3)_2]$. Since Cl^- has a greater trans effect the substitution in the monochloro intermediate $[PtCl(NH_3)_3]^+$ is most likely to occur with displacement of the ligand *trans* to the Cl^- (route i) rather than displacement of a ligand *trans* to an NH_3 ligand (route ii). *Cis*-$[PtCl_2(NH_3)_2]$ may be prepared from $[PtCl_4]^{2-}$ and NH_3.

6.6. Isomerism in Co-ordination Compounds

Isomerism

Isomerism is a general term relating to the **arrangements of ligands (or atoms within ligands)** in a compound of a specified chemical formula. Types of isomerism observed in co-ordination chemistry include: **geometrical, optical, ionisation, linkage** and **co-ordination**. For a particular chemical formula more than one type of isomerism may be observed.

Geometrical isomerism

Complexes with an identical overall gross geometry (e.g. octahedral) may be isomeric if the angles between specified ligands within the complexes are different. These are referred to as **geometrical isomers**. Square-planar $[ML_2X_2]$ and octahedral $[ML_2X_4]$ complexes can adopt such isomers in which the L ligands are at $90°$ or $180°$ to one another. These are also known as **cis ($90°$)** and **trans ($180°$)** geometrical isomers (Figure 6.6a). Such isomers are not observable in tetrahedral $[ML_2X_2]$ complexes. More complicated systems may exist e.g. *cis, cis, trans*-$[MA_2B_2C_2]$. Another type of geometrical isomerism is displayed in octahedral complexes containing two types of monodentate ligands (three of each type) i.e. $[MA_3X_3]$. Two geometrical isomeric forms are available and these are called the **mer** (meridonal) or **fac** (facial) isomers. In the *fac* isomer the ligands are all **mutually cis** and are at the vertices of one triangular **face** of the octahedron. In the *mer* isomer the ligands take up a **T-shaped** arrangement (Figure 6.6b). Again, more complicated examples of geometrical isomerism may exist e.g. *fac, cis*-$[Mn(CO)_3(PPh_3)_2Br]$.

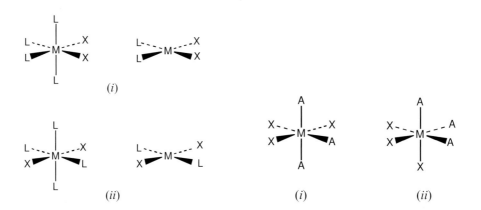

Figure 6.6a Geometrical isomers. (i) *cis* isomers in octahedral $[ML_4X_2]$ and square-planar $[ML_2X_2]$ complexes. (ii) *trans* isomers in octahedral $[ML_4X_2]$ and square-planar $[ML_2X_2]$ complexes.

Figure 6.6b Geometrical isomers of octahedral $[MA_3X_3]$ complexes. (i) The meridonal (*mer*) isomer, and (ii) the facial (*fac*) isomer.

Optical isomerism

Molecules that are **non-superimposable mirror images** are called **optical isomers**. Optical isomers will rotate a plane of polarised light in opposite directions. Optical isomerism is relatively common in organic/biological chemistry, but in inorganic chemistry it occurs less frequently. However, in transition-metal chemistry, octahedral compounds containing two (or three) bidenate ligands often display such isomerism. For complexes with two bidentate ligands, e.g. $[M(LL)_2X_2]$, the *cis* isomer is optically active but the *trans* isomer is not (Figure 6.6c). The compound $[Co(en)_3]^{3+}$, containing three bidentate ligands, is a good example of a compound which exists in two optically active forms (Figure 6.6d).

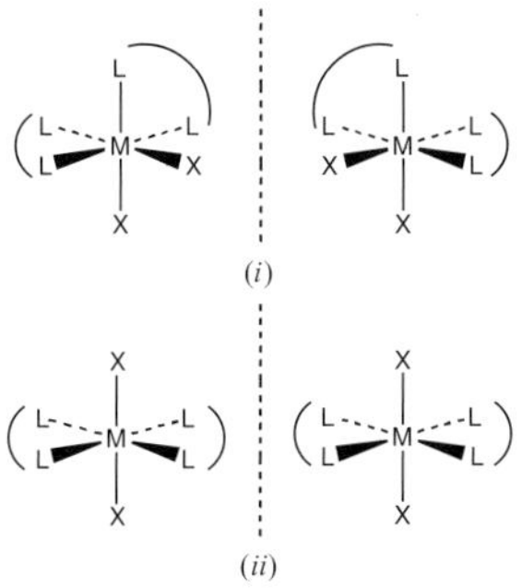

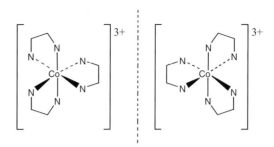

Figure 6.6c (i) Optical isomers of *cis*-[M(L$_2$)$_2$X$_2$] complexes. (ii) The *trans* isomer is not optically active as its mirror images are superimposable.

Figure 6.6d Octahedral [M(L$_2$)$_3$] complexes such as [Co(en)$_3$]$^{3+}$ can exist as two optically active non-superimposable mirror images (enantiomers).

Ionisation isomerism

Compounds with the same formula but which differ with respect to which anions are co-ordinated and which are present as counterions are described as **ionisation isomers**. These isomers yield different ions when dissolved in solution e.g. [Co(NH$_3$)$_4$Cl$_2$]NO$_2$ and [Co(NH$_3$)$_4$Cl(NO$_2$)]Cl. Both of these isomeric salts contain (different) complex cations carrying a 1+ charge, but in the former the anionic counterion is nitrite(1$-$), whereas the latter has the chloride(1$-$) anion as counterion. The following three compounds of formula 'CrCl$_3$.6H$_2$O' are also ionisation isomers: the violet [Cr(H$_2$O)$_6$]Cl$_3$, the green [Cr(H$_2$O)$_5$Cl]Cl$_2$.H$_2$O and the green [Cr(H$_2$O)$_4$Cl$_2$]Cl.2H$_2$O. In these complexes the Cl$^-$ can be analysed by quantitative AgCl formation by precipitation with excess AgNO$_3$(aq). The non-co-ordinated 'extra waters' are loosely held 'water of crystallisation' which can be removed by dehydration with concentrated H$_2$SO$_4$. Molar conductivity measurements will show how many ions are present in solution and this may help discriminate between the various ionisation isomers: values typical of 4, 3 and 2 ions would be observed for [Cr(H$_2$O)$_6$]Cl$_3$, [Cr(H$_2$O)$_5$Cl]Cl$_2$.H$_2$O and [Cr(H$_2$O)$_4$Cl$_2$]Cl.2H$_2$O, respectively.

Linkage isomerism

Some ligands may have two potential (and different) donor atoms and may co-ordinate to the metal centre through either. These ligands are described as **ambidendate** and any associated isomerism is referred to as **linkage isomerism**. Good examples are the nitrite anion, NO$_2^-$ (which is called a 'nitro' ligand when co-ordinated to a metal through N, and 'nitrito' ligand when co-ordinated through O) and the thiocyanate anion, SCN$^-$, which may co-ordinate through S or N. Sulfur dioxide, SO$_2$ is also a good example of an ambidentate ligand which may co-ordinate through O or S (examples are shown in Figure 4.9d). The donor atom of an ambidentate ligand which co-ordinates to a particular metal centre is often determined by whether the metal is hard or soft, in accordance with Pearson's hard/soft acid/base theory.

Co-ordination isomerism

In compounds where both the cation and anion are themselves co-ordination complexes the distribution of ligands between metals may vary. Such isomers are described as **co-ordination isomers**. An example of a pair of co-ordination isomers is [Co(NH$_3$)$_6$][Cr(CN)$_6$] and [Cr(NH$_3$)$_6$][Co(CN)$_6$].

6.7. Reactions of Transition-Metal Aqua Ions

Solution structure of aqua ions

Transition-metal ions in aqueous solution are invariably complexed by H_2O molecules and exist in the form of **aqua ions,** $[M(H_2O)_n]^{x+}$. Typically $n = 6$ and these aqua complexes have an octahedral arrangement of the H_2O ligands, for $Zn^{2+}(aq)$ $n = 4$ and the complex is tetrahedral. These complexes are often written as $M^{x+}(aq)$, or more simply as M^{x+}.

Reactivity of aqua ions

There are **three types of reactivity** generally displayed by aqua ions in aqueous solution: redox behaviour, acidity and ligand exchange (substitution) reactions.

Redox behaviour

Variable oxidation states are a characteristic property of transition metals and redox behaviour e.g. oxidation of $M^{2+}(aq)$ and reduction of $M^{3+}(aq)$ ions by suitable oxidants/reductants is not uncommon. Standard reduction potentials for some M^{3+} ions at pH = 0 are shown below.

Half-reaction	E°/V
$Ti^{3+} + e^- \rightleftharpoons Ti^{2+}$	−0.37
$V^{3+} + e^- \rightleftharpoons V^{2+}$	−0.26
$Cr^{3+} + e^- \rightleftharpoons Cr^{2+}$	−0.41
$Mn^{3+} + e^- \rightleftharpoons Mn^{2+}$	+ 1.51
$Fe^{3+} + e^- \rightleftharpoons Fe^{2+}$	+ 0.77
$Co^{3+} + e^- \rightleftharpoons Co^{2+}$	+ 1.82

Acidity

Aqua ions in aqueous solution are generally acidic with pK_a values dependent upon the charge and size of the hydrated ions. Values for typical ions are Ni^{2+} (8.9), Cr^{3+}(3.9), Ce^{4+}(0.1).

$$[M(H_2O)_6]^{n+} + H_2O \rightleftharpoons [M(H_2O)_5(OH)]^{(n-1)+} + H_3O^+$$

Addition of $NH_3(aq)$ or $NaOH(aq)$ to an aqueous solution of an aqua ion invariably leads, by successive deprotonation reactions, to precipitation of the hydrated hydroxide

$$[Fe(H_2O)_6]^{2+}(aq) + 2OH^-(aq) \rightleftharpoons [Fe(H_2O)_4(OH)_2](s) + 2H_2O(l)$$

However, some hydrated metal(II) hydroxides [e.g. Fe(II), Co(II), Cu(II), Zn(II)] are **amphoteric** and will dissolve in excess strong alkali to form anionic **–ate** complexes, e.g. zincate(II):

$$[Zn(H_2O)_2(OH)_2](s) + 2OH^-(aq) \rightleftharpoons [Zn(OH)_4]^{2-}(aq) + 2H_2O(l)$$

The aerial (O_2) oxidation of divalent metal ions is much easier in alkaline solution, since it is easier to lose an electron from a negatively charged or neutral complex, rather than from a positively charged one e.g. Mn(II) and Fe(II) are readily oxidised to M(III).

Trivalent metal ions are sufficiently acidic to react with alkali metal carbonates to form the hydrated metal hydroxides and CO_2:

$$2[Fe(H_2O)_6]^{3+}(aq) + 3CO_3^{2-}(aq) \rightleftharpoons 2[Fe(H_2O)_3(OH)_3](s) + 3H_2O(l) + 3CO_2(g)$$

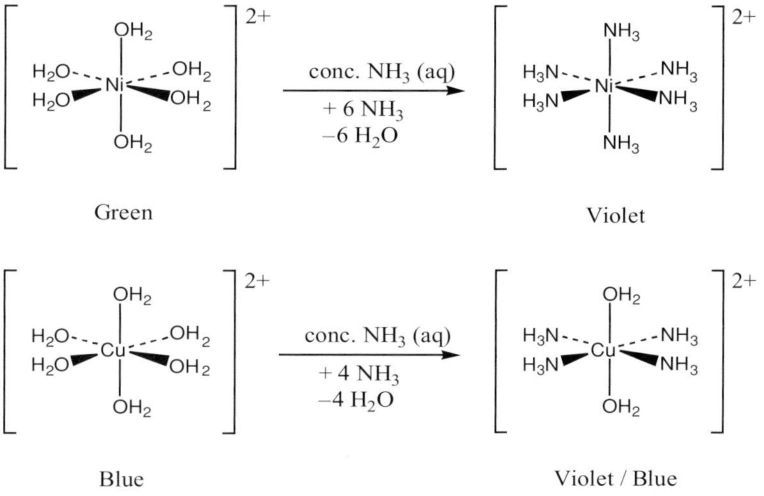

Figure 6.7a Reaction of Ni^{2+}(aq) and Cu^{2+}(aq) with conc. NH_3(aq).

Ligand substitution

Addition of excess NH_3(aq) to transition-metal aqua ion solutions ultimately leads to **ammine complexes** in which all (or occasionally some) of the H_2O ligands of the aqua complex are substituted by **NH_3 ligands**. The final product depends upon the concentration of the NH_3(aq) and values of the stepwise formation constants. The violet $[Ni(NH_3)_6]^{2+}$ is readily obtained by addition of concentrated NH_3(aq) solution to green Ni^{2+}(aq) salts, but for Cu^{2+}(aq) the violet-blue tetra(ammine) complex $[Cu(H_2O)_2(NH_3)_4]^{2+}$ is obtained (Figure 6.7a). The hexa(ammine) complex $[Cu(NH_3)_6]^{2+}$ may be obtained in liquid NH_3. Cr(III) and Co(II) also readily form 6 co-ordinate octahedral hexa(ammine) complexes, Zn(II) forms the 4 co-ordinate tetrahedral tetra(ammine) complex and Cu(I) and Ag(I) have preferred co-ordination numbers of 2 with linear di(ammine) complexes.

Addition of excess Cl^- to transition-metal aqua ions solutions leads to **chloro** complexes in which the H_2O ligands of the aqua complex are **substituted** by Cl^- **ligands**. One important point to note is that these anionic ligands often force a change in the co-ordination number and geometry of the complex with 4 co-ordinate tetrahedral complexes being the norm:

$$[Cu(H_2O)_6]^{2+} + 4Cl^- = [CuCl_4]^{2-} + 6H_2O$$
$$\text{blue} \qquad\qquad \text{yellow/green}$$

The pale violet $[Fe(H_2O)_6]^{3+}$ and pink $[Co(H_2O)_6]^{2+}$ aqua ions are converted to yellow $[FeCl_4]^-$ and blue $[CoCl_4]^{2-}$ by treatment with conc. HCl (Figure 6.7b) The larger Cl^- ligand (compared to H_2O) and its negative charge are believed to be responsible for the change in co-ordination number/geometry through steric/electrostatic interactions.

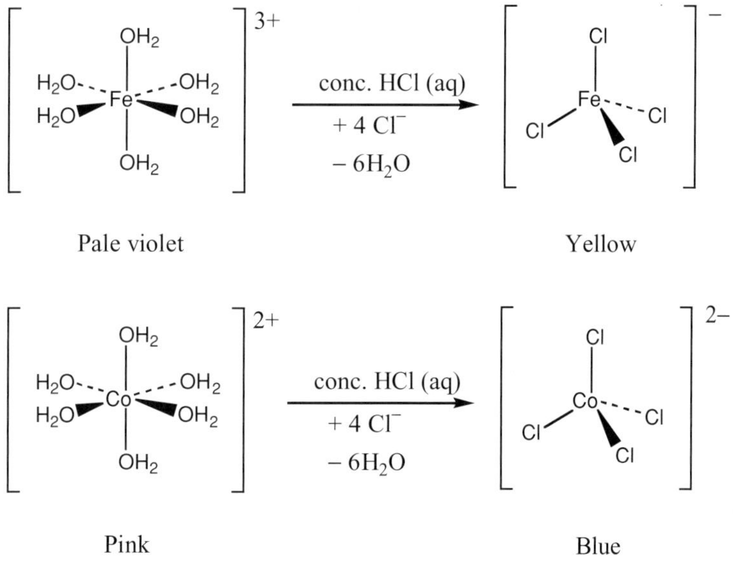

Figure 6.7b Reaction of Fe^{3+}(aq) and Co^{2+}(aq) with conc. HCl(aq).

6.8. Crystal Field and Ligand Field Theory

Introduction

Paramagnetism and colouration of many transition-metal complexes may be readily explained by use of a simple electrostatic approach to the bonding between the ligands and the metal centre. This approach is referred to as **crystal field theory**. A more detailed approach allows for covalent contributions to metal–ligand bonding and this more refined model is described as **ligand field theory**.

In crystal field theory the metal ion is considered the centre of positive charge. It is approached (surrounded) by a set of ligands in a defined geometry with each ligand considered as a point negative charge. There is therefore an **electrostatic attractive (bonding) force** between the ligands and the metal, and a **lowering of overall energy**. This is slightly offset, however, since the negatively charged electrons in the metal's valence subshell (the d-shell) feel a **repulsive force** because of the approaching negative point charges, and the **d-orbitals will be destabilised**, i.e. will increase in energy. The extent of this destabilisation on individual d-orbitals will be dependent upon their orientation with respect to ligand geometry and number. This is described in detail below for the three most common co-ordination geometries.

Octahedral complexes

We can describe the orientation of the d-orbitals as shown in Figure 6.8a, with lobes of the $d_{x^2-y^2}$ and d_{z^2} orbitals pointing along the x, y, and z axes and the lobes of the d_{xy}, d_{xz}, and d_{yz} pointing between these axes. If we were to surround the metal with a sphere of negative charge (at distance r) then all the d-orbitals would be raised in energy to the same extent. If the charge of the sphere were now concentrated as six point charges (ligands) in an octahedral arrangement along the x, y, and z axes at distance r, we can easily see that the repulsive interaction will be greatest for the d_{z^2} and $d_{x^2-y^2}$ orbitals, since these interact most strongly with the surrounding point charges. In fact, the d_{xy}, d_{xz}, and d_{yz} will now be at a lowered energy (compared to their energy in the charged sphere) as they are now no longer pointing directly at negative charge. Since there is no change in overall charge or distance there will be no change in the overall energy of the system, and the two orbitals should be raised in energy by 3/2 times as much as the three other orbitals are reduced in energy, Figure 6.8b. The energy between these two sets of orbitals is called the **crystal field splitting** and is called Δ_{oct}, Δ_o or **10Dq**. These two sets of energy levels are often referred to by symmetry labels, with the higher energy, doubly degenerate set called the e_g level and the lower energy, triply degenerate set the t_{2g} level. At a more refined level (ligand field theory) the $d_{x^2-y^2}$ and the d_{z^2} orbitals are major contributors to a degenerate pair of M−L antibonding orbitals which are given the symmetry label $e_g{}^*$.

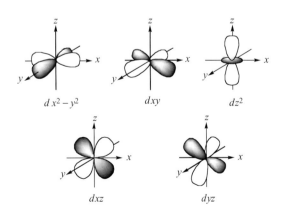

Figure 6.8a The spatial arrangement of the five d-orbitals.

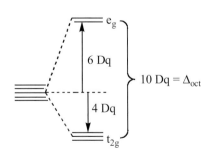

Figure 6.8b Concentrating electron density at six point charges in an octahedral field results in an altering of the relative energies of the d-orbitals.

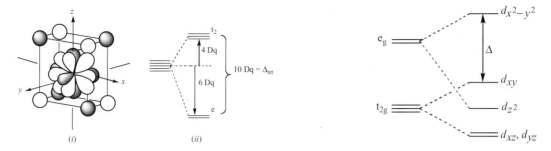

Figure 6.8c (i) The complete set of *d*-orbitals in a cubic field. (ii) The splitting of the *d*-orbitals in a tetrahedral field.

Figure 6.8d The *d*-orbital splitting in a square-planar field.

Tetrahedral complexes

A tetrahedron may be considered as derived from a cube, and if we arrange our axes so that the x, y, and z axes emerge at the centre of the faces of the cube, then none of the *d*-orbitals point directly at the negative point charges, Figure 6.8c. However, the d_{xz}, d_{yz}, and d_{xy} orbitals interact more strongly with the point charges and are destabilised relative to the $d_{x^2-y^2}$ and d_{z^2} orbitals. The **crystal field splitting** here is called $\Delta_{\mathbf{tet}}$ or $\Delta_{\mathbf{t}}$.

Square-planar complexes

Removal of two point charges along the z-axis of an octahedral geometry results in a square-planar geometry. The effect of this change on *d*-orbital energies is that those *d*-orbitals with a component in the z direction are reduced in energy, with the d_{z^2} orbital substantially reduced and the d_{xz} and d_{yz} reduced only slightly. The result is that one orbital ($d_{x^2-y^2}$) is at much higher energy than the others and the **crystal field splitting** (Δ) is defined as the energy gap between the $d_{x^2-y^2}$ and the d_{xy} orbital (Figure 6.8d). Metal complexes with a d^8 configuration often adopt this geometry as a consequence of this crystal field stabilisation.

Jahn–Teller effect

The octahedral geometry for a metal with six identical ligands is often distorted and these distortions can often be traced back to the electronic occupancy of the *d*-subshell. A well-known example is the $[Cu(H_2O)_6]^{2+}$ ion, which has a geometry with two (trans) H_2O ligands at a greater distance from the metal than the other four ligands (Figure 6.8e). The idealised octahedral structure would have had an unequal (unsymmetrical) distribution of electrons between the degenerate e_g orbitals. Distortion of the complex, in this case towards a square-planar geometry by lengthening two ligand–metal bonds, allows for the stabilisation of one of these orbitals and a net lowering in energy for the system. The distortion is an example of the so-called **Jahn–Teller** effect, and such distortions commonly occur in complexes in which degenerate orbitals are not equally occupied.

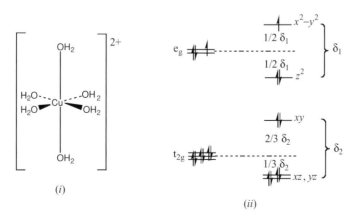

Figure 6.8e (i) As a consequence of the Jahn–Teller effect $[Cu(H_2O)_6]^{2+}$ has a distorted octahedral geometry. The effect on the *d*-orbital splitting is shown in (ii).

6.9. Ligand Field Theory and Electronic Spectra (Colours) of Transition-Metal Complexes

Colours of transition-metal complexes

The discussion in Section 6.8 demonstrated that the d-orbitals are split in energy when placed in various co-ordination geometries. The origin of colour in transition-metal complexes can often be traced back to electrons being excited from one d-orbital energy level to another. The energy of these transitions is dependent upon Δ. Such transitions are called **d–d transitions**, and in general, they are of **much weaker intensity** compared to those observed for say π–π^* transitions in organic molecules, since the former are 'forbidden' by the **Laporte** selection rule. This rule may be simply stated as 'Δl (change in orbital quantum number) for an allowed transition must be $+1$ or -1'. Since a d–d transition has $\Delta l = 0$ then it is forbidden. The observation that electronic transitions can be related to the crystal field splitting, Δ, is not a failure of the Laporte selection rule, but an indication that crystal field theory is over simplistic.

An alternative reason for colouration of transition-metal complexes, and one which can be used for any d-electron configuration, and which must be significant in d^0 and d^{10} complexes, is associated with **charge transfer (CT) bands**. Here, electron transitions between filled ligand orbitals and empty metal orbitals (LMCT) or, alternatively, filled metal orbitals to empty ligand orbitals (MLCT), are responsible for the colour. The LMCT are favoured in cases where the metal is in a high oxidation state and is readily reducible, and ligands are easily oxidisable. A good example of LMCT is the d^0 MnO_4^- anion which is an intense purple colour. Possible electronic transitions are shown in Figure 6.9a.

A third possible reason for colour in transition-metal complexes involves electronic transitions that are solely **ligand based**, and which are more or less independent of the metal. These are usually π–π^* transitions involving aromatic ligands with extended π-systems e.g. porphyrin ligands.

Factors that affect the magnitude of Δ

A number of factors combine to affect the magnitude of Δ and these are described below:

(1) As discussed in Section 6.8 the **co-ordination geometry** has a profound influence on the magnitude of Δ, e.g. keeping other factors constant $\Delta_{tet} = 4/9\Delta_{oct}$.

(2) Greater splitting is always observed when, keeping other factors constant, the **oxidation state of the metal** is increased. This can be traced back to an electrostatic interaction since the more highly positively charged metal pulls the ligands closer, enabling stronger interactions with the metal's d-orbitals.

(3) The $5d$-orbitals of transition elements from the 3rd transition-series are much more diffuse and extend further out towards the ligands than the $3d$-orbitals of transition elements of the 1st series. The electrostatic d-orbital/ligand interactions are therefore stronger, and keeping other factors constant, **descending a Group** causes an increase in the crystal field splitting.

(4) The **electronic properties of the ligand** influence the magnitude of Δ. It is possible to arrange ligands in order of increasing crystal field splitting, and this is described as the **spectrochemical series**. The spectrochemical series is not quite as one would expect from considering purely electrostatic ligand/metal interactions, and some degree of metal/ligand **covalency** (particularly π-bonding) needs to be taken into account. This refinement of crystal field splitting is termed **ligand field splitting**. The π-acceptor ligands, e.g. CN^-, are termed 'high field' ligands, whilst π-donor ligands, e.g. I^-, are termed 'weak field' ligands.

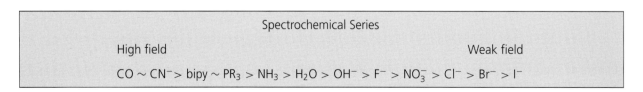

Spectrochemical Series	
High field	Weak field
$CO \sim CN^- >$ bipy $\sim PR_3 > NH_3 > H_2O > OH^- > F^- > NO_3^- > Cl^- > Br^- > I^-$	

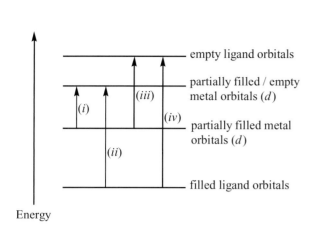

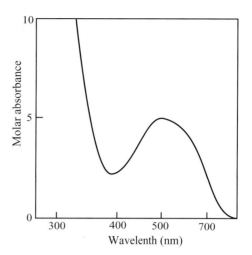

Figure 6.9a Examples of possible electronic transitions which might be observed in a transition-metal complex. (i) Metal based d–d transitions, (ii) ligand–metal charge transfer bands (LMCT), (iii) metal–ligand charge transfer bands (MLCT), and (iv) ligand based transitions. Charge transfer and ligand based transitions, are 'allowed' and are usually much more intense than 'Laporte-forbidden' d–d transitions.

Figure 6.9b The electronic spectrum of $[Ti(H_2O)_6]^{3+}$ showing a d–d transition with λ_{max} at 510 nm.

The spectrochemical series gives an indication of the expected magnitude of Δ and hence is a means of predicting where to expect the maximum absorption in the visible spectrum of the complex (λ_{max}). This might be considered further by using two Ti^{3+} (d^1) metal complexes as examples: $[Ti(H_2O)_6]^{3+}$ has λ_{max} at 510 nm (Figure 6.9b) and $[TiBr_6]^{3-}$ has λ_{max} at 850 nm. As the Br^- is a weaker field ligand than H_2O we would expect the energy gap between the t_{2g} and e_g orbital (Δ) to diminish in size (Figure 6.9c). Since $E = hc/\lambda$ and the energy gap is now smaller, λ_{max} must increase on changing H_2O for Br^-.

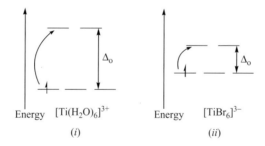

Figure 6.9c The d-orbital splittings of $[Ti(H_2O)_6]^{3+}$ and $[TiBr_6]^{3-}$ indicating the effect of the ligand in the spectrochemical series.

6.10. Ligand Field Theory and Magnetic Properties

Overview

The number of d-electrons, and their arrangement within the d-orbitals of the transition-metal share responsibility for the magnetic properties of the metal in its compounds. In particular, complexes with one or more unpaired electrons display paramagnetic (or ferromagnetic) properties, whereas complexes in which all the electrons are paired are diamagnetic. The magnetic properties of compounds may be investigated quantitatively by magnetic susceptibility experiments.

Magnetic susceptibility

It is possible to measure the magnetic susceptibility of a material by use of a magnetic balance e.g. a Gouy balance. The magnet is an electromagnet. The balance weighs a known volume of the material in the presence and in the absence of a strong magnetic field (Figure 6.10a). A paramagnetic material will be drawn into the magnetic field when the magnet is on, whereas a diamagnetic material is weakly repelled. The observed change in weight for the sample is an aggregate of any paramagnetic properties (due to unpaired electrons) and diamagnetic effects (which are common to all matter). Usually the paramagnetic term is much larger than the diamagnetic term, and the latter is often ignored. However, for highly accurate measurements diamagnetic 'correction' factors are used. From the data obtained experimentally, it is possible to calculate the **molar susceptibility** (χ_m) of the sample and from this (ignoring the small diamagnetic contribution) it is possible to calculate, using a simplified version of Curie's equation, the **magnetic moment** (μ) (in units of Bohr magnetons) of the paramagnetic species present. If χ_m is positive the sample is paramagnetic; if χ_m is negative it is diamagnetic.

$$\mu = 2.84(\chi_m.T)^{1/2}$$

The magnetic moment of an atom is usually composed of an electron-spin angular momentum component and an orbital angular momentum component. In transition-metal complexes of the **1st transition-series** the magnetic moment is generally associated with electron-spin angular momentum only (the orbital angular momentum is quenched) and this is called **spin-only paramagnetism**. It is therefore possible to predict the magnetic moment (in units of Bohr magnetons) of a paramagnetic ion, on the basis of the number of unpaired electrons present (n) by use of the **spin-only formula**:

$$\mu = [n(n + 2)]^{1/2}$$

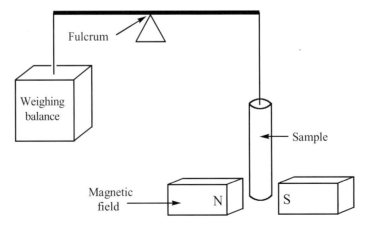

Figure 6.10a A schematic diagram of a Gouy balance. For convenience the magnet is electromagnetic, and the sample is weighed in the presence, and absence of the magnetic field.

Thus, by measuring the magnetic susceptibility of a sample and comparing the observed magnetic moment with predicted values obtained by the spin-only formula, it is possible to determine how many unpaired electrons there are present in the transition-metal complex. Some examples of experimentally determined and calculated values are given in Figure 6.10b.

Ion	Experimental value (μ/BM)	No. of unpaired electrons (n)	'Spin-only' value (μ/BM)
Ti^{3+}	1.7–1.8	1	1.73
V^{3+}	2.7–2.9	2	2.83
Cr^{3+}	~3.8	3	3.87
Mn^{3+}	~4.8	4	4.90
Fe^{3+}	~5.9	5	5.92

Figure 6.10b Experimental and calculated (spin-only formula) values of μ/BM for selected metal ions.

High-spin and low-spin complexes

The splitting of the d-orbitals in octahedral, square-planar and tetrahedral crystal field geometries was examined in Section 6.8. In an octahedral crystal field, two alternative electronic configurations are possible for each of the d^4, d^5, d^6 and d^7 ions. These alternatives **differ by the number of unpaired electrons** within the ions and are called the high-spin and low-spin configurations. For example, the d^4 Cr^{2+} ion may have an electronic configuration of $(t_{2g})^4(e_g)^0$ with two unpaired electrons (low-spin), or $(t_{2g})^3(e_g)^1$ with four unpaired electrons (high-spin). The lowest energy configuration will be the preferred alternative and this will depend upon the relative magnitudes of the **crystal field splitting (Δ_o)** and any energy required to force electrons to pair in the same orbital, the **pairing energy, P**. In cases where $\Delta_o < P$ then high-spin complexes are observed, and where $\Delta_o > P$ low-spin complexes are observed, (Figure 6.10c). The Cr^{2+} (aq) ion, with relatively weak field H_2O ligands, has a high-spin configuration indicating that $\Delta_o < P$. In general, high-spin octahedral complexes are the norm for 1st series transition-elements, unless π-acceptor ligands are present, or the metal is in a high oxidation state e.g. Co^{3+}. Tetrahedral complexes are normally high-spin ($\Delta_{tet} = 4/9\Delta_{oct}$) and square-planar complexes are normally low-spin. Because larger crystal field splitting is more commonly observed for 2nd and 3rd series transition-elements, their complexes are invariably low-spin.

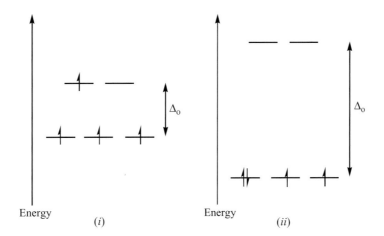

Figure 6.10c Energy level diagrams showing (i) high-spin $(t_{2g})^3(e_g)^1$ and (ii) low-spin $(t_{2g})^4(e_g)^0$ electronic configurations for the d^4 Cr^{2+} ion. In the high-spin case, $\Delta_{oct} < P$ (pairing energy) whereas in the low-spin case $\Delta_{oct} > P$.

Ferromagnetism

Magnetic interactions between paramagnetic centres that are in close proximity (as in bulk metals) or separated by small linking atoms (e.g. oxides, fluorides) may occur. If the magnetic dipoles are **aligned in the same direction** in a larger domain, then the **paramagnetism is greatly enhanced**, and this phenomenon is termed **ferromagnetism**. Such behaviour is commonly observed in Fe, Ni, Co, magnetite (Fe_3O_4), γ-Fe_2O_3, CrO_2; the latter three are used in magnetic recording tapes.

6.11. Metal Carbonyl and Organometallic Chemistry

Organometallic compounds

Organometallic compounds are compounds with **metal to carbon bonding**. The bonding may be through σ-bonds or π-bonds, or a combination of both, and typical ligands are carbon monoxide (CO), isocyanides (CNR) and hydrocarbons. These hydrocarbon ligands range from simple alkyl groups, with M–C σ-bonds, to π-donor ligands such as $CH_2=CH_2$, η^6-C_6H_6. With π-bonding ligands the number of carbon atoms (n) bound to the metal centre is specified in the notation η^n. Compounds such as $[Cr(\eta^6\text{-}C_6H_6)_2]$ and $[Fe(\eta^5\text{-}C_5H_5)_2]$ are referred to as **sandwich compounds** because the metal is sandwiched between the hydrocarbon ligands. Half-sandwich compounds e.g. $[Re(CO)_3(\eta^5\text{-}C_5H_5)]$ are also known.

The 18-electron rule

The number of electrons within the valence shell of the metal in organometallic transition-metal compounds has a profound influence on the compounds' stability/reactivity. In particular, compounds with 18 electrons within this shell are **usually stable** and such compounds are said to obey the **'18-electron rule'**. The requirement for 18 electrons arises from the fact that the metal possesses nine valence orbitals $\{5\,nd + 3\,(n+1)p + 1\,(n+1)s\}$ and once these are filled the metal has the electronic configuration of an inert gas. The '18-electron rule' is sometimes referred to as the 'noble gas rule' or the 'effective atomic number rule'. To arrive at an electron count the number of valence electrons on the neutral metal is added to the number of electrons donated to the metal from its associated neutral ligands, and finally any residual charge on the complex must also be accounted for. Electron counting examples are shown in Figure 6.11a. The 'electron donor numbers' of ligands commonly found in organometallic chemistry are:

1 electron donors:	F, Cl, Br, I, (pseudo-halogens), H, alkyl/aryl groups
2 electron donors:	H_2O, NH_3, PR_3, SR_2, CO, η^2-$CH_2=CH_2$
3 electron donors:	η^3-C_3H_5 (allyl)
4 electron donors:	η^4-C_4H_6 (butadiene)
5 electron donors:	η^5-C_5H_5 (cyclopentadiene, Cp)
6 electron donors:	η^6-C_6H_6 (benzene)

Some transition-metal complexes are stable with fewer than 18 electrons in their valence shell but it is very rare to exceed this number. In particular, **16-electron metal-centres** are common for low-spin d^8 ions in a **square-planar co-ordination** geometry, e.g. Pt^{2+}, and in some complexes involving elements from the left hand side of the d-block e.g. $[Ti(\eta^5\text{-}C_5H_5)_2Cl_2]$.

	(i) **[Fe(η^5-C$_5$H$_5$)$_2$]**	(ii) *cis*-**[PtCl$_2$(NH$_3$)$_2$]**	(iii) **[Cr(η^6-C$_6$H$_6$)(CO)$_3$]**.
Metal's valence electron count	8	10	6
Electron count from ligands	10 from 5 x 2 C$_5$H$_5$	6 from 1 x 2 Cl = 2 and 2 x 2 NH$_3$ = 4	12 from 6 x 1 C$_6$H$_6$ = 6 and 2 x 3 CO = 6
TOTAL	18	16	18

Figure 6.11a Examples of electron counting as applied to the 18e$^-$ rule. (i) [Fe(η^5-C$_5$H$_5$)$_2$], (ii) *cis*-[PtCl$_2$(NH$_3$)$_2$], (iii) [Cr(η^6-C$_6$H$_6$)(CO)$_3$].

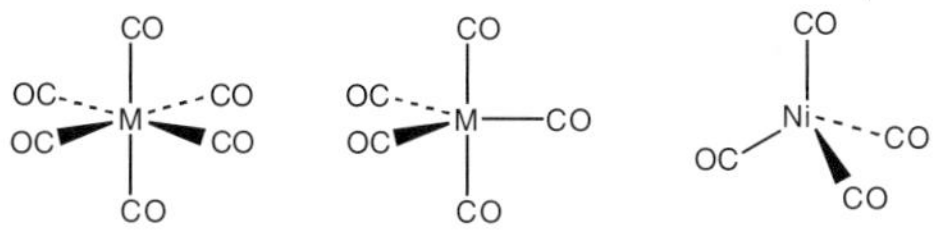

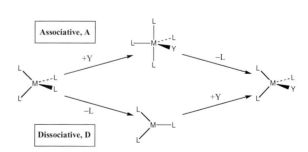

(i) (ii) (iii) (i) (ii)

Figure 6.11b Examples of structures of mononuclear metal carbonyl complexes. (i) [M(CO)$_6$] (M = Cr, Mo, W), (ii) [M(CO)$_5$] (M = Fe, Ru) and (iii) [Ni(CO)$_4$].

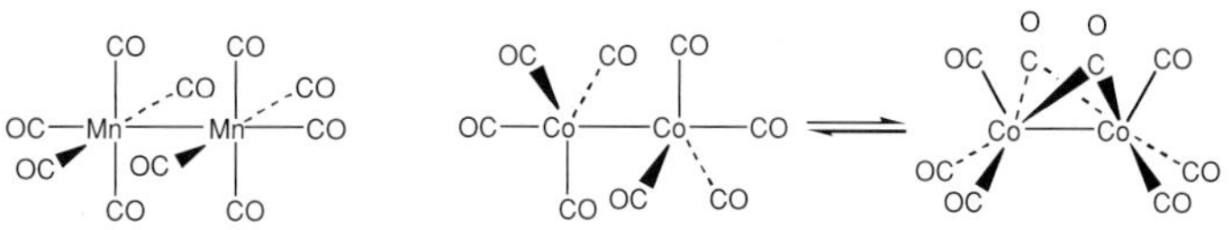

Figure 6.11c Two examples of some dinuclear metal carbonyl complexes with metal–metal bonds. (i) [Mn$_2$(CO)$_{10}$], (ii) isomers of [Co$_2$(CO)$_8$].

Structures of binary metal carbonyl species

Complexes of metals with only carbonyl groups as ligands are described as **binary metal carbonyl** species. These complexes generally follow the 18-electron rule and metals with an 'even' number of valence electrons can easily satisfy that rule by combining with the appropriate number of CO ligands to form **mononuclear species**. Examples are: [Cr(CO)$_6$], [Mo(CO)$_6$], [W(CO)$_6$], [Fe(CO)$_5$], [Ru(CO)$_5$], [Ni(CO)$_4$] (Figure 6.11b).

Metals with 'odd' electron valence electron counts require the formation of metal–metal bonds (a metal can be considered as a 1 electron donor) in addition to carbonyl ligands to achieve stable 18-electron systems, and **dinuclear species** result. Examples are: [Mn$_2$(CO)$_{10}$], [Co$_2$(CO)$_8$]. Carbonyl ligands may bridge two (or more) metal atoms and their electron pair must be shared accordingly in each of the metal's electron count to be compliant with the 18-electron rule (Figure 6.11c).

Dinuclear and **polynuclear** species are also obtained from mononuclear derivatives by elimination of CO and formation of metal–metal bonds. Thus, the 18-electron rule about each metal centre is maintained in the following condensed species: [M$_2$(CO)$_9$] (M = Fe, Os), [M$_3$(CO)$_{12}$] (M = Fe, Ru, Os) and [M$_4$(CO)$_{12}$] (M = Co, Rh, Ir). Larger polynuclear metal carbonyl clusters exist but their structures are best considered in terms of delocalised bonding. Many of these structures are illustrated in Figure 6.11d.

Substitution mechanisms

Mechanistic studies of ligand substitution reactions of organometallic transition-metal complexes indicate that reactions broadly follow one of two mechanistic pathways. Although this is illustrated for a tetrahedral complex in Figure 6.11e, these reaction pathways are also applicable to octahedral and square-planar complexes.

Since 20-electron species are very rare, 18-electron complexes tend to undergo ligand substitution by firstly eliminating a ligand (the leaving group) to give a 16-electron intermediate, followed by addition of the new ligand. The intermediate has a co-ordination number one fewer than the reactant or product. This process is termed a **dissociative mechanism** and given the symbol **D**, and is very common in organometallic chemistry.

Alternatively, complexes with 16 or fewer electrons may add the incoming ligand first and then eliminate the leaving group (Figure 6.11e). In this mechanism the intermediate has two electrons more at the metal centre than either reactant or product, and the process is called an **associative mechanism** and given the symbol **A**. The intermediate in this mechanism has one more ligand than either the reactant or product.

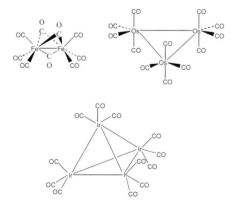

Figure 6.11d Some examples of 'condensed' dinuclear and polynuclear metal carbonyl species.

Figure 6.11e Two possible reaction mechanisms (A and D) for ligand substitution in a tetrahedral complex.

6.12. *d*-Block Elements and Industrial Chemistry

Production of Fe and steel

Reduction of iron ore (Fe_2O_3) to Fe metal is a very important industrial process and is the most important application of the **coke pyrometallurgy industry**. Traditionally, the reaction is carried out in a **blast furnace** which is loaded with iron ore (hematite, Fe_2O_3), coke (C), and limestone ($CaCO_3$). Air is injected through the bottom of the furnace and, at the operating temperature (2000°C), the coke is transformed into CO which is the reducing agent and which is ultimately oxidised to CO_2. The Fe_2O_3 is first reduced to Fe_3O_4 (magnetite), then FeO (wustite) and finally Fe. The molten Fe, which is rich in C (~4%) and other impurities (P, Si, Mn, S) collects at the bottom of the blast furnace (Figure 6.12a). It is run off to cool and form **pig-iron**. Pig-iron is hard but very brittle. The action of heat on the limestone transforms it to lime (CaO) which reacts with impurities in the ore (mainly silicates) to form calcium silicate. The calcium silicate collects above the molten iron and is removed as molten slag.

Pig-iron was traditionally converted to **wrought-iron** by the **Bessemer** process, in which more hematite was mixed with the molten pig-iron. Air was blown through the mixture and the excess C and other impurities burnt out. Wrought-iron is ideal for mechanical working, being both tough and malleable.

Nowadays however, the bulk of the pig-iron is converted directly into **steel**, with C content of 0.3–1.5% and few other elemental impurities. This is accomplished by blowing a high presure O_2 jet through a molten agitated mixture of slag/pig-iron to oxidise impurities from the Fe into the slag, which is then run-off (**basic oxygen process**). Steel is stronger and more workable than pig-iron, or wrought-iron. Stainless steel is resistant to corrosion and is an alloy of Fe and Cr (~10%).

Production of TiO₂

TiO_2 is a non-toxic, unreactive, opaque solid and is commonly used as the white pigment in **paint**. It is also used in the paper, plastics and ceramics industries. The majority of TiO_2 is obtained from its ore ilmenite ($FeTiO_3$) by the **sulfuric acid** process. The ore is dissolved in H_2SO_4 and the sulfates $FeSO_4$ and $TiOSO_4$ are leached into solution. Concentration of the solution under vacuum yields crystalline $FeSO_4.7H_2O$ which is removed by centrifugation. Further concentration of the

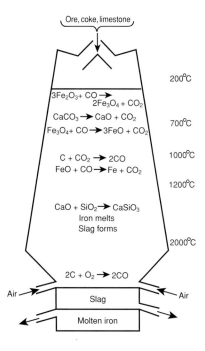

Figure 6.12a Diagram of a blast furnace.

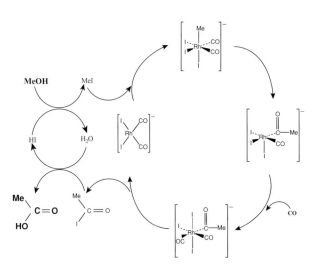

Figure 6.12b Reaction cycles in the synthesis of ethanoic acid via the Monsanto process.

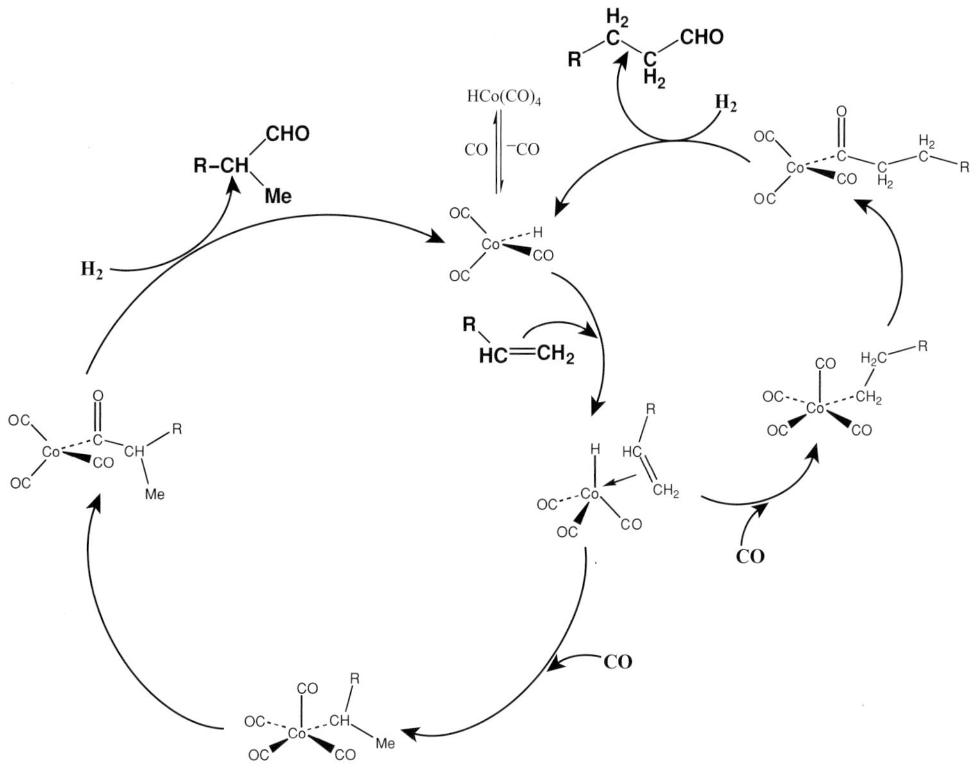

Figure 6.12c Two competitive cycles observed in the catalytic hydroformylation of alkenes (Oxo process). The cycle on the right yields linear aldehydes, whilst that on the left affords branched aldehydes.

solution followed by heat treatment hydrolyses the $TiOSO_4$ to insoluble $TiO_2.H_2O$. Finally the hydrate is heated at $900°C$ to TiO_2.

Purification of nickel

Purification of Ni used to be accomplished by the **Mond** process. This process takes advantage of the fact that $[Ni(CO)_4]$ is formed reversibly and rapidly from CO gas and elemental Ni at low pressures and temperatures ($50°C$). The $[Ni(CO)_4]$ is volatile and is thermally decomposed at higher temperatures ($200°C$) to high purity Ni and CO. The compound $[Ni(CO)_4]$ is extremely toxic. Nowadays, purification of Ni is mostly accomplished electrochemically.

Homogeneous catalysis

Many important organic transformations are catalysed by transition-metal complexes, and two examples are given here. The **Monsanto process** is used on a large scale for the conversion of MeOH to $MeCO_2H$ by addition of CO. The reaction is catalysed by the unsaturated 16-electron complex $[Rh(CO)_2I_2]^-$ (Figure 6.12b). The cobalt carbonyl hydride complex $[HCo(CO)_4]$ catalyses the conversion of alkenes to aldehydes, by addition of CO and H_2, in the hydroformylation (**Oxo process**) reaction (Figure 6.12c).

Heterogeneous catalysis

Many industrially important processes are catalysed by transition-metals themselves or their binary compounds; some of these are described in detail in other sections of this book. The **Sasol** and the **Fischer–Tropsch** processes both generate hydrocarbons from synthesis gas (H_2 + CO) and the usual catalysts are Co or Fe. The **Ziegler–Natta** system (Section 4.13) uses $TiCl_4/AlEt_3$ and is very important in the polymer industry for the polymerisation of alkenes. The **contact** process uses V_2O_5 to catalyse the oxidation of SO_2 to SO_3 (Section 4.10). Iron will catalyse the production of NH_3 from N_2 and H_2 in the **Haber** process (Section 4.14). Most automobiles are now fitted with 'catalytic converters' which contain active Pt, Pd and Rh particles dispersed over an Al_2O_3 coating of a honeycomb ceramic. **Catalytic converters** convert all petrol emission products into 'acceptable' products, CO_2, N_2 and H_2O.

f-Block Elements – Lanthanides and Actinides

7.1. The 4f Elements – the Lanthanides or Rare Earths

Introduction

The lanthanides are a series of 14 elements (La–Yb) which occur immediately following Ba ($Z = 56$) in the Periodic Table. Their chemical properties are very similar and the generic symbol Ln is often used.

Electronic configurations

For the gaseous atoms these are generally $4f^n 6s^2$. As a result of the similar energies of the $4f$, $5d$ and $6s$ orbitals in this region of the Periodic Table there are some exceptions: La [Xe] $5d^1 6s^2$, Ce [Xe] $4f^1 5d^1 6s^2$ and Gd [Xe] $4f^7 5d^1 6s^2$.

The trivalent ions, the most chemically important for all Ln, have configurations [Xe]$4f^n$. In the ions the $4f$ orbitals have very **limited spatial extension** and cannot effectively reach the periphery of the atom to covalently interact with counterions or neutral ligands. Thus, in the vast majority of their compounds and complexes the **bonding is electrostatic in nature**.

Lanthanide contraction

Across any row of the Periodic Table the increase in **effective nuclear charge** causes atoms and ions to become physically smaller. For the 14 lanthanides this effect is particularly pronounced, and is termed the **lanthanide contraction**. For instance the 6-co-ordinate radii decrease from 117 pm (La^{3+}) to 100 pm (Yb^{3+}). This has important consequences for the chemistry of the lanthanides (and also for the following $5d$ and $6p$ elements, see Sections 4.1 and 6.1). The effects on the structure of lanthanide compounds are well documented. For instance, in the solid state $LnCl_3$ the co-ordination number of Ln is 9 for La–Gd, and 6 for Tb–Yb.

Oxidation states

The trivalent oxidation state

For all Ln **the +3** oxidation state is the most stable. For some there is a limited chemistry in the +4 and +2 oxidation states. The range of oxidation states is summarised in Figure 7.1a. The reason for this preference for the +3 oxidation state lies in the magnitudes of the 3rd and 4th ionisation energies, I(3) and I(4). Where I(4) is low the +4 oxidation state is more accessible, and where I(3) is high the Ln(II) state will be encountered. The variation of I(4) and I(3) for the lanthanides is shown in Figure 7.1b and correlates well with the observed +4 and +2 oxidation states, respectively. Other factors must, however, also be considered and analysis using thermochemical cycles gives a more detailed interpretation of the observed trends. Mainly as a result of the lack of covalent bonding, the range of oxidation states available to the lanthanides is very limited. Contrast this behaviour with d-block metals and early actinides where the range of oxidation states is much greater due to the stabilising effect of covalent bonding.

La	Ce	Pr	Nd	Pm	Sm	Eu	Gd	Tb	Dy	Ho	Er	Tm	Yb
	+4	+4*	+4*					+4	+4*				
+3	+3	+3	+3	+3	+3	+3	+3	+3	+3	+3	+3	+3	+3
					+2	+2			+2*			+2	+2

Figure 7.1a The oxidation states of the lanthanides – the most stable oxidation state is underlined. *only observed in the solid-state.

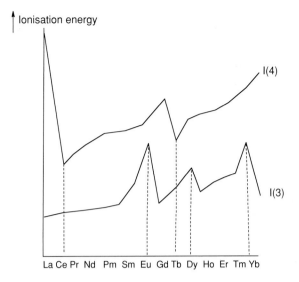

Figure 7.1b A schematic representation of the 3rd and 4th ionisation energies of $Ln(g)^{n+}$ and their correlation with the stability of Ln^{4+} and Ln^{2+} ions.

The +2 and +4 oxidation states

They are mostly only stable in the solid-state, either oxidising (Ln^{4+}) or reducing (Ln^{2+}) water.

$$2Ln^{4+} + H_2O \rightarrow 2Ln^{3+} + \tfrac{1}{2}O_2 + 2H^+$$
$$2Ln^{2+} + 2H_3O^+ \rightarrow 2Ln^{3+} + H_2 + 2H_2O$$

Some kinetic stability is observed for Ce(IV) and Eu(II) and aqueous solutions of these can be stable.

Typical chemistry

The metals are prepared from the oxides or halides by high temperature ($1000°C$) reduction with more electropositive metals.

$$Ln_2O_3 + 3Ca \rightarrow 2Ln + 3CaO$$
$$2LnX_3 + 3Ca \rightarrow 2Ln + 3CaX_2$$

The metals are highly reducing and react readily with air and water:

$$2Ln + 6H_2O \rightarrow 2Ln(OH)_3 + 3H_2$$
$$4Ln + 3O_2 \rightarrow 2Ln_2O_3$$

The properties of the compounds are as expected from large, redox inactive ions and a wide range of common salts are known, such as LnX_3 (X = F, Cl, Br, I), $Ln_2(SO_4)_3$, $Ln(NO_3)_3$, $Ln_2(CO_3)_3$, etc.

Co-ordination complexes are formed principally with oxygen donors where the bonding interaction is ion–ion or ion–dipole. Geometries of co-ordination compounds do not conform to any simple system and are generally distorted from idealised geometries. Co-ordination numbers tend to be high as a result of the large ionic radii.

7.2. The 5f Elements – the Actinides

Occurrence

The generic symbol An is frequently used for the actinides. Only thorium (Th), protactinium (Pa) and uranium (U) occur naturally on Earth in workable amounts. Neptunium (Np) and plutonium (Pu) occur as traces in uranium ores. The **transuranic elements** (those following uranium) have all been synthesised by nuclear processes. Figure 7.2a shows the conversion of uranium to plutonium; the ^{238}U nucleus absorbs a neutron and emits a gamma ray. Successive β-emission leads to plutonium. The most readily available isotopes of the actinides are listed in Figure 7.2b. It is worth noting here that the isotope listed is not necessarily the most stable. For example ^{244}Pu ($t_{\frac{1}{2}} = 8.2 \times 10^7$ y) is much more stable than ^{239}Pu, but as ^{239}Pu is readily formed by neutron bombardment of ^{238}U it is available in much greater quantities.

$$^{238}U + n \longrightarrow {}^{239}U + \gamma$$
$$^{239}U - \beta^- \longrightarrow {}^{239}Np$$
$$^{239}Np - \beta^- \longrightarrow {}^{239}Pu$$

Figure 7.2a   Nuclear synthesis of ^{239}Pu from ^{238}U.

Electronic configuration

In this region of the Periodic Table the $5f$, $6d$ and $7s$ orbitals have similar energies. This is reflected in the electronic configuration of the gaseous actinide atoms, which are shown in Figure 7.2b. For the gaseous ions the picture simplifies as the $5f$ becomes more stable than the $6d$- and $7s$-orbitals, and a configuration $[Rn]5f^n$ is found for all the elements.

Element	Symbol	Half life	Amount obtainable	5f	6d	7s
Actinium	^{227}Ac	21.8 y	milligrams		1	2
Thorium	^{232}Th	1.39x10^{10} y	multikilograms		2	2
Protactinium	^{231}Pa	3.25x10^4 y	multikilograms	2	1	2
Uranium	^{238}U	4.51x10^9 y	multikilograms	3	1	2
Neptunium	^{237}Np	2.14x10^6 y	multikilograms	4	1	2
Plutonium	^{239}Pu	24400 y	multikilograms	6		2
Americium	^{243}Am	7370 y	100 g	7		2
Curium	^{244}Cm	17.6 y	milligrams	7	1	2
Berkelium	^{249}Bk	341d	milligrams	9		2
Californium	^{252}Cf	2.65 y	micrograms	10		2
Einsteinium	^{254}Es	276 d	micrograms	11		2
Fermium	^{257}Fm	80 d	unweighable	12		2
Mendelevium	^{258}Md	53 d	unweighable	13		2
Nobelium	^{255}No	3 min	unweighable	14		2

Figure 7.2b   Availability of actinides and electronic configuration of their gaseous atoms.

Typical chemistry

The **oxidation states** of the actinides are shown below, with the most stable in aqueous solution underlined. The $5f$-orbitals extend further from the nucleus than the $4f$ in the lanthanides, and thus there is the possibility of covalent overlap. Similarly, even though the $6d$ and $7s$ lie at higher energy than the $4f$ in the ions, they are sufficiently close in energy to be accessible and hence used in bonding.

Ac	Th	Pa	U	Np	Pu	Am	Cm	Bk	Cf	Es	Fm	Md	No
	2					2			2	2	2	2	**2**
3	3	3	3	3	3	**3**	**3**	**3**	**3**	**3**	**3**	**3**	3
	4	4	**4**	4	**4**	4	4	4	4				
		5	5	**5**	5	5							
			6	6	6	6							
				7	7								

The wide range of oxidation states available for the early actinides is in sharp contrast to the range found in the lanthanides, and is reminiscent of the *d*-block oxidation states and indicates **significant covalent bonding** in many of their compounds. Note that from americium onwards there is an increasing stability of the +3 oxidation state – i.e. a more lanthanide-like character. This is manifested by the **increasing oxidising ability** of the higher oxidation states, thus whilst U(VI) in UO_2^{2+} is relatively poorly oxidising the americium analogue is a strong oxidising agent. Other differences from the lanthanides are also seen in the later actinides with the availability of the +2 oxidation state in solution and with evidence to suggest that No^{2+} is the most stable oxidation state for that element, as a result of the $5f^{14}$ configuration. The chemistry of the actinides has been extensively studied due to their applications as nuclear fuels and their subsequent reprocessing. The volatile and highly reactive UF_6 is of considerable technological importance in isotopically enriching uranium. Its synthesis from uranium dioxide is shown in Figure 7.2c. The UO_2 is first converted to UF_4 which is oxidised with ClF_3. The difficulty in oxidation reflects the instability of UF_6 and highlights how the nature of the groups bonded to the metal critically influences the stability. Both UF_6 and **the linear uranyl ion, UO_2^{2+}** contain uranium(VI), the uranyl ion is stable and a poor oxidant whilst the hexafluoride is unstable and powerfully oxidising. The difference in reactivity is probably due to the relative π-bonding abilities of oxygen and fluorine, with the formal double bond of oxygen increasing the electron density at the metal.

$$UO_2 + 4HF \longrightarrow UF_4 + 2H_2O$$
$$3UF_4 + 2ClF_3 \longrightarrow 3UF_6 + Cl_2$$

Figure 7.2c The production of UF_6.

The **Latimer diagrams** for uranium and plutonium (Figure 7.2d) show that the metals are highly electropositive and will react with water (Figure 1.5a), and that UO_2^+ and PuO_2^+ should both **disproportionate**. The similarity in the E^0 values involving PuO_2^{2+}, PuO_2^+, Pu^{4+} and Pu^{3+} means that there is a complex equilibrium between all Pu oxidation states in aqueous solution; a delicate balance that can be strongly influenced by **complexing agents** and is further complicated by radiological effects which destabilise high oxidation states for ^{239}Pu by producing a reducing environment. The Latimer diagram for americium is discussed in Section 1.5.

$$UO_2^{2+} \xrightarrow{0.06\ V} UO_2^+ \xrightarrow{0.58\ V} U^{4+} \xrightarrow{-0.63\ V} U^{3+} \xrightarrow{-1.70\ V} U$$

$$PuO_2^{2+} \xrightarrow{0.92\ V} PuO_2^+ \xrightarrow{1.17\ V} Pu^{4+} \xrightarrow{0.99\ V} Pu^{3+} \xrightarrow{-2.03\ V} Pu$$

Figure 7.2d Latimer diagrams for uranium and plutonium in acidic solution.

7.3. Some Applications of f-Block Elements

Lanthanides

Medical imaging

Gadolinium complexes such as Gd DTPA (Figure 7.3a) find extensive use in magnetic resonance imaging (MRI). The presence of seven unpaired electrons on the Gd^{3+} ion ($4f^7$ configuration) means that species containing Gd(III) are highly **paramagnetic** (Section 6.10) and thus reduce the NMR relaxation times of nuclei close to them. For application in MRI the Gd complex is selectively absorbed into one type of tissue and the ^1H nuclei in the water molecules relax more rapidly than those in the surrounding areas and thus provide an enhanced contrast. The formation of a **stable chelate complex** reduces the toxicity of Gd as it is bound strongly to the ligand and does not enter the body as the free $Gd^{3+}(aq)$ ion.

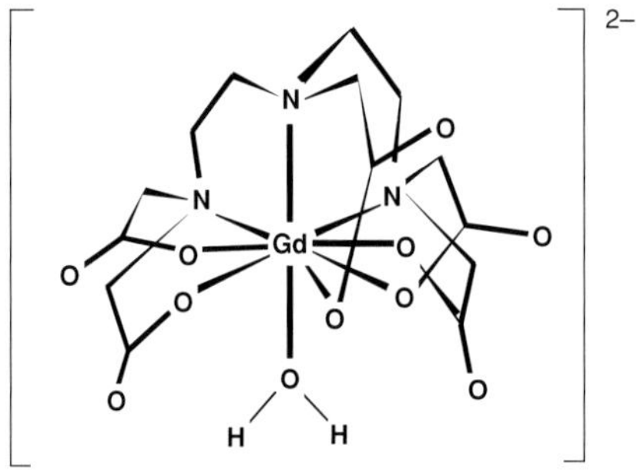

Figure 7.3a The structure of the Gd-DTPA complex used in magnetic resonance imaging (DTPA = diethylenetriaminepentaacetic acid).

Catalytic converters

There is considerable interest in the use of Ce in catalytic converters where the relatively inexpensive cerium provides an alternative to platinum metal based catalysts. The catalysis is based on the oxidation of CO by CeO_2, the Ce_2O_3 being reoxidised to Ce(IV) by air. The reactions involved are shown in Figure 7.3b.

$$Ce_2O_3 + \tfrac{1}{2}O_2 \longrightarrow 2CeO_2$$

$$2CeO_2 + CO \longrightarrow Ce_2O_3 + CO_2$$

Figure 7.3b The catalytic oxidation of CO by oxygen using a Ce catalyst.

Actinides

Nuclear fuels reprocessing

The major application of actinides is in **nuclear energy** production and the associated treatment of spent fuel. The **reprocessing** of fuel requires the separation of the fission products (La and Tc) from the reusable uranium and plutonium, followed by the separation of the uranium from the plutonium. The reprocessing relies on the differences between the species in the solubilities of their compounds and the strength of the ligand to metal bonding in their co-ordination complexes. The Ln^{3+} and An^{3+} ions tend to form weak complexes which readily dissociate, whilst ions with higher oxidation state metals such as An^{4+} and AnO_2^{2+} bind more strongly to ligands. The general order of stability of complexes of the actinides is $AnO_2^{2+} > An^{4+} > An^{3+}$. The actinide oxidation states become more oxidising on moving across the series and therefore for any given oxidation state plutonium will be more oxidising than uranium.

100

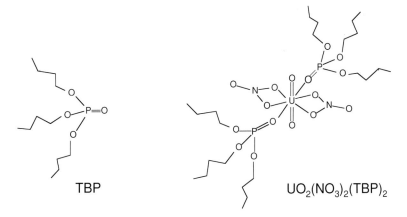

Figure 7.3c The structure of tributylphosphate (TBP) and its complex with uranyl nitrate.

Reprocessing in the **THORP process** (thermal oxide reprocessing) involves three stages:

1. separation of the uranium and plutonium from the other fission products,
2. separation of the uranium from the plutonium, and finally
3. processing the uranium and plutonium into useful forms.

Fuel rods are dissolved in 7M HNO_3 and the resulting solution is allowed to 'cool' for about 5 y so that the intensely radioactive, short-lived isotopes decompose. At this point the solution contains mainly UO_2^{2+}, Pu^{4+}, La^{3+}, TcO_4^- and Ba^{2+}. Extraction with tributylphosphate (TBP) removes the uranium and plutonium as these form strong complexes which are soluble in petrol. The extraction into the organic phase is favoured by the high aqueous concentration of nitrate, which tends to promote the formation of **electrically neutral** species based on $UO_2(NO_3)_2$ and $Pu(NO_3)_4$, and the fact that the TBP contains **hydrophobic** butyl groups. Hence neutral hydrophobic complexes are formed which are soluble in petroleum-based solvents. The structures of tributylphosphate and its complex with uranyl nitrate are shown in Figure 7.3c. Thus the fission products remain in the aqueous phase whilst the uranium and plutonium are distributed into the petroleum.

The separation of U from Pu is by **selective reduction** of Pu^{4+} to Pu^{3+}, which is readily achieved as Pu(IV) is more oxidising than U(VI). This reduction is achieved by equilibration of the U/Pu loaded organic phase with a weakly acidic solution of U^{4+} as indicated in Figure 7.3d. The lower concentration of nitrate in the aqueous solution at this stage favours the right hand side of the equilibrium. Uranium(IV) reduces Pu^{4+} as shown by the reduction potentials below and will not reduce the uranyl ion.

$$UO_2^{2+}/U^{4+} \quad E^0 = 0.27 \text{ V} \; ; \; Pu^{4+}/Pu^{3+} \quad E^0 = 0.98 \text{ V}$$
$$2Pu^{4+} + U^{4+} + 2H_2O \quad \rightarrow \quad 2Pu^{3+} + UO_2^{2+} + 4H^+ \quad E^0_{cell} = 0.71 \text{ V}$$

The Pu^{3+} thus formed is not strongly bound to the TBP and is readily extracted back into the aqueous phase, thus achieving the separation of the uranium and plutonium. The choice of U^{4+} as the reducing agent here is an ingenious one as any other would give rise to by-products which might complicate the process.

$$2Pu(NO_3)_4(TBP)_2(org) + U^{4+}(aq) + 2H_2O \rightleftharpoons 2Pu^{3+}(aq) + 6NO_3^- + UO_2(NO_3)_2(TBP)_2(org) + 2TBP(org) + 4H^+(aq)$$

Figure 7.3d The reduction of Pu(IV) by U(IV) and its partition from the organic to aqueous phase.

Smoke detectors

Many homes these days have smoke detectors. These have ^{241}Am sources as the active component. This is an α-emitter and ionises air producing small currents in the detector. When smoke enters this region the current changes and this change sets off the alarm.

Suggested Further Reading

F.A. Cotton, G. Wilkinson and P.L. Gaus, *Basic Inorganic Chemistry*, 3rd ed., John Wiley and Sons, New York (1995).

F.A. Cotton, G. Wilkinson, C.A. Murillo and M. Bochmann, *Advanced Inorganic Chemistry*, 6th ed., John Wiley and Sons, New York (1999).

S.A. Cotton, *Lanthanide and Actinide Chemistry*, John Wiley and Sons, Chichester, UK (2006).

P.A. Cox, *Instant Notes in Inorganic Chemistry*, BIOS Scientific publishers, Oxford, UK (2000).

J. Emsley, *The Elements*, 3rd ed., Oxford University Press, UK (1998).

N.N. Greenwood and A. Earnshaw, *Chemistry of the Elements*, 2nd ed., Pergamon Press, Oxford, UK (1997).

C.E. Housecroft and A.G. Sharpe, *Inorganic Chemistry*, Prentice Hall, UK (2001).

A.G. Massey, *Main Group Chemistry*, 2nd ed., John Wiley and Sons, Chichester, UK (2000).

S.M. Owen and A.T. Brooker, *A Guide to Modern Inorganic Chemistry*, Longman Scientific, Harlow, UK (1991).

D.F. Shriver and P.W. Atkins, *Inorganic Chemistry*, 3rd ed., Oxford University Press, Oxford, UK (1999).

L. Smart and E. Moore, *Solid State Chemistry*, Chapman and Hall, London, UK (1992).

T.W. Swaddle, *Applied Inorganic Chemistry*, University of Calgary Press, Calgary, Canada (1990).

Index